Illusions of Human Thinking

Gabriel Vacariu

Illusions of Human Thinking

On Concepts of Mind, Reality, and Universe in Psychology, Neuroscience, and Physics

Springer

Gabriel Vacariu
University of Bucharest
Bukarest, Romania

Copyright: Previously published in romanian language under the title: " Lumi epistemologic diferite – Noua paradigma a gandirii".

ISBN 978-3-658-10443-6 ISBN 978-3-658-10444-3 (eBook)
DOI 10.1007/978-3-658-10444-3

Library of Congress Control Number: 2015949629

Springer

Lektorat: Dr. Reinald Klockenbusch

Printed on acid-free paper

Springer is a brand of Springer Fachmedien Wiesbaden
Springer Fachmedien Wiesbaden is part of Springer Science+Business Media
(www.springer.com)

„The greatest enemy of knowledge is not ignorance, but the illusion of knowledge.“
Stephen Hawking

„Everything we call real is made of things that cannot be regarded as real.“
Niels Bohr

„I am now convinced that theoretical physics is actual philosophy.“
Max Born

„A new scientific truth does not triumph by convincing its opponents and making them see the light, but rather because its opponents eventually die, and a new generation grows up that is familiar with it.“
Max Planck

Table of Contents

Introduction

The World, Man's Greatest Illusion

In this book, I will attempt to prove that the greatest illusion about human knowledge that has existed from the dawn of time is the notion of the world, the 'uni-verse', or, as I called it, 'the unicorn world'. The world (the unicorn world) is a wrong notion just because the human being included all different types of entities in the same framework (eventually, the same spatio-temporal framework): macro-particles (like tables, stones and planets), microparticles (photons, electrons and quarks) and waves, neuronal patterns of activation, brain, bodies and mind, mental states, the self, etc. As I will show in this book, the introduction of all these types of entities within the same ontological framework is the main mistake in the history of human thinking. The notion of the world/universe is a fantasy created by human mind which wrongly inserted all different types of entities (that really exist) within the same spatiotemporal framework and this is the reason I called the world as being the 'unicorn world'.[1] I say it is the greatest illusion because human beings have built their entire knowledge within the house of the world and nobody ever questioned it.[2] More exactly, the main mistake that led to the creation of the unicorn world is that we, human beings, believe (consciously or not) that we are the only observers of the world. As a result, we have placed our Gods, as well as all other beings (including not only humans, with their minds, brains and bodies, but also plants, cells, microbes, animals) and all other objects (tables, stones, planets, electrons, waves and fields) within the same world – the unique world, the universe. This world is composed of all the existent entities and their relationships, which are inevitably placed within the same spatiotemporal framework.

The common perception we have grown used to is that all living organisms observe or perceive more or less the same world. However, nobody has ever wondered if a planet or an electron observes (as we will see, observation is

1 The unicorn is the fantastic animal, usually a horse with a horn, with some features that belong to different real animals. I call the world as the 'unicorn world' exactly for the same reason: different types of entities (that really exist) are situated, incorrectly, by human being within the same (spatiotemporal) framework.

2 Even the physicists who proposed the existence of „many-worlds" or multiverse have not rejected the unicorn-world.

equivalent to interaction) the same unique world. We will come to understand in the pages that follow that we, human beings, do not observe the world, and this is the main reason why the unicorn world does not exist. Sometimes, we recognize the fact that through perceptual mechanisms, human beings do not observe the real world: we perceive only phenomena[3]. Our perception of the world is a huge approximation of the real world (the thing-in-itself), or we can say that we perceive only certain aspects of the real world.

In the following pages, I will strive to demonstrate that the unicorn world is the greatest enemy of human knowledge, especially when it comes to its foundations. It does not have a great effect on our knowledge of daily life, nor even on the many types of knowledge acquired by certain branches of science, but it deeply affects the foundation of all knowledge we have about the world. Actually, the state of affairs concerning the foundations of human knowledge is very problematic: some of life's greatest issues remain unsolved. Some of them appeared a few centuries ago (such as the mind-body problem), others several decades ago (for example, quantum mechanic's non-locality or entanglement) and yet others only three decades ago (like many of the issues of cognitive (neuro)science). All of these problems that appeared in various essential academic domains are still unsolved, since human knowledge is restricted to its illusive home in the unicorn world, which is a construct created by human beings. This framework did allow some of the great theories of several branches of science to evolve, but it also led to great anomalies and paradoxes in the realm of human knowledge. Obviously, these anomalies and contradictions have not been solved, because humans were unable to leave the framework of the unicorn world due to the astonishing fact that our ignorance has been so profound that we could not realize we were building our entire knowledge within the wrong framework.

All the debates presented in this book are strongly related to the distinction between ontology and epistemology. When we speak of something ontologically, what we mean is that the various things discussed are independent of our tools of investigation. As a sub-domain of philosophy, ontology deals with things that really exist in the world. Human beings discover what objects/processes really exist by using tools of observation (here I refer both to our eyes and to other organs of perception, as well as to instruments such as microscopes or telescopes which we use in our observations). The knowledge we acquire by using various tools is the topic of epistemology (a sub-discipline of philosophy).

3 In Kant's philosophy, phenomena are the result of interactions between our transcendental mechanisms (pure intuitions of space and time, plus categories) and the noumena (or thing-in-itself, i.e. something that really exists independent of our perception. Transcendental means something that is beyond any experience. The problem is that even if we change our conditions of observation/perception, we will never be able to observe the thing-in-itself.

In other words, ontology deals with that which exists, while epistemology deals with the parts of that which exists which we are able to know. Ever since the Antiquity, philosophers have made a strong distinction between ontology and epistemology which, as we will see, is one of the greatest mistakes in human thinking and it is also the essence of the unicorn-world. This long-perpetuated distinction between ontology and epistemology was probably imposed by religious perspectives, which sought to reflect the limits of human knowledge as opposed to the infinite power and knowledge of Gods. Plato officially instituted the distinction that has framed human thought to this very day,[4] and the later great debates or approaches of philosophy were strongly influenced by the Platonic approach, going as far as to be variations on the same theme, which can be seen even in the works of those who tried to contradict Plato. In Kant's work, we can find the essential noumena-phenomena distinction, which is a reiteration of Plato's famous distinction between ideas (which only Gods can perceive) and appearances (which are perceived by humans, who are limited beings). It is impossible not to notice that this distinction was a very important problem in Kant's transcendental philosophy.

To this, we can add that many other Ptolemaic epicycles (by which we refer to wrong notions constructed with the wrong arguments) were created by the majority of philosophers within the unicorn-world. After all, the philosophy of the last two centuries was strongly influenced by Kant. Unfortunately, even if many philosophers tried to replace parts of Kantian philosophy, they all continued to work inside the unicorn-world. Within this paradigm (or better said, this meta-paradigm), the results of human thinking have been paradoxical. In certain branches of science, humankind has made huge progress, mainly regarding certain local knowledge connected to limited aspects of the world, but the same type of progress has definitely not been made regarding our knowledge of the entire world.

In general, when somebody proposed a scientific theory which tried to describe the whole world, that theory turned out to be wrong. While we can speak of progress in science, it has been always a contradiction in terms to promote progress in philosophy. As a rule, a systematic philosophy has to 'change the world', that is, to offer a new image of the world, a new Weltanschauung. Normally, until now nobody wondered about the world, but only about our image (imperfect, of course) of the world. Both scientists and philosophers have generally accepted the ontological distinction between reality and appearances. In science, for example, we have the case of Bohr, who accepted the Kantian distinction between noumena and phenomena. A great number of other physicists have searched for the elementary particles that compose our world. As a remnant of religious thinking, philosophers and scientists recognized the ontological and epistemological limitations of humans,

4 We have to remember that Whitehead (one of the greatest philosophers of the 20th Century) noticed, correctly, that all philosophy after Plato is a footnote of his philosophical system!

which some thinkers explained through the concept of God. Considering these limitations, it was compulsory for us to create the notion of the phenomenological world, which is the apparent world.

Amazingly, the effect of recognizing our limitations was exactly opposite to the status which was officially imposed on us: we, imperfect beings, ended up creating the world. Thus, our imperfection acquired a special status: our imperfect knowledge became 'ontologically loaded' (Kant – see Parvu 2004) because we have been dealing with the 'phenomenal world' (in other words, with appearances) and not with the 'thing-in-itself' (the 'real world' that we believe we have no access to). However, we will see in this book that the 'real world' does not exist. While philosophers have always tried to discover the relationship between the real and the phenomenal world, scientists have generally never wondered about it. Their inquiries were restricted to the objects and processes of the world. Obviously, various humans (priests, philosophers, scientists etc.) have greatly emphasized their amazing capacity of creating the world. God has been placed somewhere in a corner of the world and many human beings moved into the roles of supervisors of this world. Instead of assuming the position of limited entities, some of us became dictators of human knowledge. Inevitably, within this framework of human thought, we can find many debates or battles in the history of human knowledge, some of which had disastrous consequences.

In this work, I will try to incorporate the main ideas from my previous works (2005, 2008, 2010, 2011, 2012) in a general framework, that of epistemologically different worlds (EDWs), and I will describe its applications in various branches of science. The main idea is that the world does not exist, but what really exist are EDWs. Epistemologically different worlds are sets formed of epistemologically different objects/processes and the interactions between them. An object exists because it interacts with other objects in the same epistemological world (EW), so each EW has its own set of epistemologically different objects. An object exists only for the objects it interacts with from the same EW. Epistemologically different interactions refer to the relationships between epistemologically different objects. A particular case of this type of relationships are physical forces. We can make an (incorrect) analogy between interactions and perceptions: a table exists for me because I see it in front of me; that table exists for the coffee mug which stands (interacts due to the gravity between the table and the coffee mug) on the table precisely because that mug perceives the table it is standing on (and the table perceives the mug in its turn). (A systematic presentation of the EDWs, in the first three chapters of this book.) We have to reject the notion of the world, as well as the distinction between ontology and epistemology. Moreover, even if many ideas from Kantian philosophy are still important, we have to eradicate some of the core notions/ideas of his transcendental philosophy. One of the main revisions will concern to the noumenal-phenomenal distinction.

We have to accept biologist Konrad Lorenz's idea: during the evolution of the species, the development of our perceptual tools has been in accordance with our external macroscopic environment. That is why we can perceive or know certain real, existent characteristics of the world and not just appearances, or the phenomenal world. (Lorenz 1941) In this way, our perception gains an ontological status: we perceive the true macroscopic world. From the perspective of EDWs, this vision of perception can be extended from humans to all other entities (living and nonliving) that exist in EDWs. Each entity can observe or interact with entities of the same class in the same conditions and, therefore, each class represents an epistemological world (EW). Through the EDWs perspective, we have to reject the distinction between ontology and epistemology and to replace the world with the EDWs.

In order to show that the world does not exist, I will investigate the eternal philosophical questions (questions that science has undertaken in the last century): what exists and what are the relationships between the existent entities?. It would seem that the majority of philosophers have forgotten these questions, but as we will see, there have been both scientists and philosophers who have provided answers to them. However, because they did so within the unicorn world, those answers are at least partly wrong. The main exceptions are Einstein's theories of special and general relativity, the latter of which, for instance, correctly explains the relationships between macroscopic entities. From my point of view, the theory of relativity (in both its special and general forms) is the only great scientific theory which is correct, but it lacks the ontological foundation which uses the EDWs perspective. As opposed to this situation, quantum mechanics for instance incorporated some essential but very problematic concepts (such as non-locality or non-spatiality, or the concept of entanglement), while cognitive neuroscience is still in a „pre-history period“[5] (Kuhn), mainly because of the problematic relationship between neuronal and conceptual levels.

As much as possible I will attempt to provide some answers to the greatest problems of science within a scientific framework, not a philosophical one.[6] I am interested in defining, by using a new 'meta-paradigm' (Friedman 2001), what existed, exists or will exist in the past, present and future without the contribution of man. In other words, I reject Kant's main assumption concerning the division between noumena and phenomena. As I wrote above, some important notions (more precisely, a part of the framework) of Kantian philosophy will be extended

5 Using Thomas Kuhn's expression, it means that cognitive neuroscience is not a real science yet. (See Chapter 6 of this book)

6 However, following Max Born's belief („I am now convinced that theoretical physics is actual philosophy“), I know that the greatest problems of particular sciences (physics, biology, cognitive (neuro)science, etc.) are philosophical problems!

from humans to all entities (living and nonliving). Human beings will no longer be the only class of entities who ontologize their epistemology; instead, this property will be shared by all 'epistemologically different entities' through their interactions, which are equivalent to human perception. The point of view chosen in the constitution (to use Kant's term) of EDWs does not belong only to human beings; instead, all other entities have the same right of constituting the other entities that belong to the same EW. We cannot talk about seeing with 'God's eye' or about a 'view from nowhere': 'God's eye' belongs to the church, while the 'view from nowhere' (see Thomas Nagel, for instance) implies the existence of an empty world. Both views are meaningless in an academic environment.

Unfortunately (or maybe fortunately), we need to abandon many of the notions regulated by the ancient paradigm of the unicorn world, with its main ideas about the world or the universe, about the essence of the world, or the causality between many entities situated in the same unicorn world that presupposes a single spatio-temporal framework. This is because the majority of philosophers and scientists have been working within the unicorn world, which means that their theories are flawed: they consider that all existent entities are incorporated within the unicorn world, which creates problems mainly in the perception of the relationships (causalities) between those entities.

Nevertheless, many great philosophers who worked within the confines of the unicorn world became aware that certain problems could not be solved. This becomes immediately apparent when we look at various situations that arose in what we refer to as 'modern philosophy' (the philosophical systems developed between the 17 and the 19th centuries): Descartes desperately searched for the relationship between the mind and the brain/body; Spinoza was forced to introduce the existence of a totally unknown neutral substance, of which mind and matter were just attributes; Hume denied the notion of causality, as a reply to Newton's inability to explain the nature of gravity; Leibniz created the concept of monads which mirror the world; finally, introducing the noumenal-phenomenal distinction, Kant tried to develop the philosophical foundations that would support Newton's theory. All these presuppositions were fabricated within the unicorn world in answer to the many pseudo-problems that dominated philosophy. The perspective of EDWs will go beyond these (and other) alternatives. However, I imported something from each of these approaches for the construction of the EDWs and within the current book I will elaborate the axiomatic-hyperontological framework (which contains 11 propositions) available for every known or unknown EW.

If we reject the hegemony of man, who is confined to certain conditions when he observes the 'unique world', we have to wonder about the conditions of observation/perception available to other entities. In my framework, I will offer a detailed demonstration of the fact that the conditions of observation are equivalent to the conditions of interaction. The question which arises in this context is 'Does every entity observe the same world?'. As we will see below, each class of entities represents an EW that is epistemologically different from other EWs, which means that we have many worlds, but these worlds are neither ontological (forming the many-worlds or multiverse), nor epistemological aspects of the same world. Instead, they are epistemologically different worlds, which consist of epistemologically different entities and their interactions. As I will argue in this book, the epistemologically different viewpoints/interactions have the same ontological status of objectivity: all share the same objective reality. In other words, all EDWs share the same objective reality.

I am certain that many people will consider this project either too ambitious or hopeless. The question 'What exists?' is either too complicated (and therefore nobody can answer it, so philosophers are forced to limit themselves to analyzing concepts or to moral-political philosophy), or it is considered old-fashioned (which means that searching for an answer is worthless). We have to take into account that, in the last century, nobody even tried to create a Weltanschauung.

I think that this state of affairs is not paradoxical for at least two reasons: (1) The framework continues to be the unicorn world; (2) Science absorbed almost all the particular problems of old philosophy. In other words, scientists have been trying to explain all the phenomena in the world (entities and processes, relationships and functions etc.), with each branch of science occupying a place in describing a part of this unique world. In the last two centuries, the trend has been to insert more and more mathematics in such descriptions of the world (even this tendency is a remnant of Plato's framework).[7] Furthermore, the language of different scientific fields (and of their subdivisions) became more and more specialized. As a result, the philosopher, who once had to accumulate the essential knowledge of different branches of science, which were not as numerous as they are today, was completely overcome by the actual scientific knowledge that explains the world.

Under these circumstances, the philosopher abandoned the struggle to create a Weltanschauung and retired with solemn dignity under the logical-linguistic cover of analyzing linguistic notions[8], and later got involved in complex moral and political debates. However, the powerlessness of the field is reflected by the attitude

7 About the extreme mathematization of theories in physics (the superstring theory), see Vacariu and Vacariu (2010).

8 If logical positivism (the starting point of analytic philosophy at the beginning of the 20th century) was a reply to the absurd metaphysics of the final part of the 19th century, nevertheless, the metamorphosis of a tool (the analysis of language) into a goal shows the incapacity of philosophers to create a new Weltanschauung of the world. This incapacity mirrors the extraordinary disputes among the scientists of the last 100 years in their great efforts to explain the unicorn world. Within such a disputable framework it was indeed impossible for a philosopher to generate a new philosophy.

of the majority of philosophers in rejecting even the temptation to create a Weltanschauung. Devastated by the enormous amount of knowledge produced by various branches of science, philosophers lost themselves in sterile discussions on profound topics, being overcome by the decisive steps forward taken by great scientists.

We have to remember that a long time ago philosophy and science were strongly interconnected, or better said, they were united in a paradigm (the Weltanschauung) that could dominate human thought for a short or even for a long period. Long ago, the great scientist was a philosopher, while the great philosopher was a scientist. After being under the control of philosophy for a long time, science was finally liberated from dogmatic and dictatorial[9] philosophical paradigms (I am referring here to the religious framework which proved unsatisfactory a long time ago). Today, scientists are right to ignore contemporary philosophers completely, but this does not mean that a lot of philosophers do not continue to try, hopelessly, to talk about the philosophy of science in general, or about the philosophy of a particular science, but their endeavor is usually pitiable, since they only provide certain 'analytical textbooks' regarding some notions of a particular or general science, which fail to arouse the interest of scientists.

The relationship between philosophy and science has completely broken down, so the question we are confronted with is: How can we develop a new philosophy, which will influence science decisively? The physicists of the last century rejected their contemporary philosophy. They considered that „philosophy today is dead" (Hawking and Mlodinow, 2010[10]). I believe that in this case physicists are right, since contemporary philosophers have nothing to do with the elaboration of a Weltanschauung which the various branches of science would find necessary, since all of them (physics, biology, cognitive science etc.) are confronted with great problems, which continue to be unsolved after many decades of research.

9 A philosophical approach has to be a dictatorial way of thinking, otherwise it is not philosophy, or better said, it is not pure philosophy. Its dictatorship involves the change of an old paradigm with a new one.

10 This book starts with the following paragraph: „We each exist for but a short time, and in that time explore but a small part of the whole universe. But humans are a curious species. We wonder, we seek answers. Living in this vast world that is by turns kind and cruel, and gazing at the immense heavens above, people have always asked a multitude of questions: How can we understand the world in which we find ourselves? How does the universe behave? What is the nature of reality? Where did all this come from? Did the universe need a creator? Most of us do not spend most of our time worrying about these questions, but almost all of us worry about them some of the time. Traditionally these are questions for philosophy, but philosophy is dead. Philosophy has not kept up with modern developments in science, particularly physics. Scientists have become the bearers of the torch of discovery in our quest for knowledge." (Hawking and Mlodinow 2010, p. 13) With my EDWs perspective, I offered a completely new answer to these questions and I showed that the answers given by scientists to these questions have been wrong.

Scientists continue to struggle with these issues today, but they do not expect any help from philosophers due to the great break between science and philosophy which took place at least 150 years ago.

The main issues which appeared in separate fields of science were created by people (scientists, philosophers etc.) who worked within the framework created by the belief in the existence of the world/universe or, as I called it, the 'unicorn world'. I will furnish a new Weltanschauung that transforms many such scientific or philosophic problems into pseudo-problems. My message is mainly for scientists, not for philosophers. My EDWs perspective is completely different from any kind of philosophy that appeared in the history of philosophy. According to Ilie Parvu (during a private conversation), a philosophical 'image of the world' (Weltanschauung) is created by a philosopher immediately before or after a great theory in science.

If this image appears before a scientific theory, it implies the existence of many paradoxes/anomalies (Thomas Kuhn) that are the results of working within an old philosophical or scientific paradigm. If it appears afterward, it means that the philosophical foundations of that scientific theory are absent and the philosopher takes on the job of providing them (for example Kant, who responded to Newton). In both cases, we can say that scientists provide theories of local knowledge about particular phenomena. The missing element is a framework that incorporates the global knowledge representing the paradigm within which the majority of scientists usually work. Scientists are usually correct in ignoring the (meta)paradigm in which they work or think, since they investigate specific phenomena that do not explicitly include the paradigm. They manipulate concepts that can be empirical or not and which are not essential to the meta-paradigm within which they function. They do not try to furnish an 'image of the world', but only an image of a part of the world. Nevertheless, even working within these confines, scientists are unable to solve local problems completely, since it would involve using certain fundamental concepts.

The unicorn world has been, in general, quite peaceful. Even if it is a fictitious paradigm, it allows the progress of various scientific fields. However, when knowledge reaches certain limits, the unicorn makes its appearance more and more violently. The researcher or philosopher is surprised, then astonished and finally outraged by certain unexplained phenomena. After some time, this leads to a better paradigm successfully replacing the old one. Paradoxically, this paradigm shift creates even deeper problems, since our knowledge moves closer and closer to the non-existing foundations of the unicorn world. This context leads to very strange situations: entities in general exist, but nobody realizes that the general framework in which they are included is wrong.

Analyzing the truly great problems of (certain branches of) science, I believe that both the scientist and the philosopher need to return to the old relationship between philosophy and science. Following Friedman, (with his 'meta-paradigms') (Friedman 2001), we can now return, with the aid of my new approach, the EDWs perspective, to the „long forgotten image of philosophy that once guided science" (Friedman 2001). This movement is more necessary than ever because in the last hundred years scientific researchers and philosophers worked in an ocean of knowledge overfilled and clogged by many deep anomalies and mysteries. This change would mean the replacement of the unicorn world with EDWs.

One of the major problems that appears in both philosophy and science is the relationship between entities. I want to emphasize that, from an EDWs perspective, the relationships between entities that belong to separate EDWs are only of correspondence, not of causality. More precisely, questions concerning possible links of causality between two EDWs (or between entities that belong to two EDWs) are meaningless and to consider their existence would mean thinking in the wrong framework, because this would mean committing the fundamental error of placing them within the same spatio-temporal framework of the unicorn world. Such relationships are meaningless within the framework of EDWs, since this error leads to major problems.

Thus, even in this section, i shift the focus of Hume's doubt about causality (Newton's gravitational force acting on macro-objects) towards the causality between various entities that belong to different EDWs (such as the mind and the brain, waves and particles, or microparticles and macroparticles). The relationships between such entities have created the greatest problems in human knowledge, many of which were unfortunately pseudo-problems that were impossible to solve within the given framework. They have either been explained through the introduction of incredible Ptolemaic epicycles, or they have remained unexplained. Hume would be delighted by the endless disputes regarding the relationship between the mind and the brain, waves and particles, or microparticles and macroparticles.

For instance, extremely competent physicists have tried to find the relationship between Einstein's theory and quantum mechanics, or the one between waves and particles, and the scientist who explains these relationships will probably be awarded the Nobel prize. Even the relationship between the mind and the brain is no longer a topic reserved for philosophers, as many scientists in the realm of cognitive neuroscience are trying to explain the relationship between the two entities. From my viewpoint, these relationships do not exist. As already stated, the main cause for this way of thinking is that these objects (entities) really do exist, but they are placed in EDWs, not in the unicorn world. The scientific theories which refer to entities belonging to the same EW are generally correct. It is the theories about the relationship between entities that belong to EDWs that led to huge errors.[11]

I deeply believe that nature is much simpler than we have thought. If Hawking wrote that „the greatest enemy of knowledge is not ignorance, but the illusion of knowledge“, then the unicorn world is the greatest enemy of knowledge. Consequently, it will be extremely difficult to change man's mentality, in which the unicorn world is very well-established. In this work, through the axiomatic-hyperontological framework available for any EW, we finally return to pure philosophy (Weltanschauung). From the EDWs perspective, the Weltanschauung does not depend on human perception/conception. The EDWs perspective is beyond Kant's noumena-phenomena distinction. The notion of noumena is meaningless, the world has to be replaced with EDWs.

What is essential is that the notion of observation is equivalent with that of interaction. Imagine that you, the reader, are a microparticle (an electron or a photon) or a macroparticle (a table or a planet; a cell or a multicellular organism). Obviously, an electron does not observe as such, but it interacts with something. What is this something? An electron interacts with/observes other microparticles from the same EW. An electron does not interact with a planet. Given this equivalence, I will introduce the following rule, which rejects the dictatorial status of man as the sole observer: Humans are not the only entities who observe (interact with) other entities.

As we have seen in the first part of this introduction, many paradoxes, anomalies and contradictions in science and philosophy appeared simply because people broke this rule. Usually, when a notion is successfully used in a theory which explains phenomena belonging to a particular EW, it is incorporated in other theories with the hope of explaining the phenomena belonging to other EDWs. Obviously, this dangerous extrapolation has been possible because of the framework of the unicorn world.

Nevertheless, even in the paradigm of the unicorn world, Kant elaborated the same rule. Explaining the difference between objective validity and objective reality in Kant's philosophy, Hanna comments on some paragraphs from Critique of Pure Reason (A239/B298-9 and A248/B305), writing that „empty concepts cannot be meaningfully applied by us either to noumenal objects or to objects of our sensory intuition, and in that sense they are impossible- that is, impossible to use“. (Hanna 2001, p. 90-1) In the middle of the last century, Carnap introduced a similar rule with his linguistic frameworks. (Carnap 1950)

11 In the last chapters of the book, I will offer two examples of such errors: the essential notions of entanglement or non-locality from quantum theory and Searle's idea (1992) that the brain produces the mind (we can find this idea in Frith's 2007 work). It must be mentioned here that Searle is philosopher, while Frith is a scientist working in cognitive neuroscience.

While Kant was partly accurate in introducing this rule when it comes to language ('one language', since people were unaware of the existence of micro-particles at that time; partly, because Kant avoided the mind-body problem), Carnap uses the notion of 'empty concepts' after the great debates between Einstein and researchers working in quantum mechanics. Carnap's movement (1950) is just a philosophical recoil from ontology to logic and language. With Wittgenstein (one of the most important philosophers at the beginning of the last century), the analysis of language became the main occupation for philosophers instead of dealing with science and systematic philosophy. ('Systematic philosophy' refers to the philosophical approaches that furnish a Weltanschauung.)

From my point of view, we have to be aware of our ontological/epistemological limits when we structure our knowledge of what really exists. The only way we can do this is by trying to abide by the rule of avoiding those concepts which are impossible to use. Paradoxically, the philosophers of the last hundred years, who worked in the area of analytic philosophy, made the major mistake of breaking the Kant-Carnap linguistic rule. This infringement reflects the difference between ontology (what exists - the subject of pure philosophy) and language (inquires about various notions – the topic of analytic philosophy).

In the final part of this introduction, I want to place the emphasis again on my belief about the essential step a philosopher must take in order to create a pure philosophy akin to those created by great philosophers such as Plato, Aristotle, Kant, Hegel etc. The philosopher needs to investigate the main problems of science (and of its particular branches), and then to go beyond specific scientific frameworks (that contain specific notions). The mission of a philosopher is not to elaborate empty notions (the philosophical achievements of the last century) but to furnish a new framework of thinking for scientists who make inquiries into the fundamental problems of nature.[12] This new framework has to fit the correct scientific theories perfectly, to clarify or reject problematic scientific concepts, and to reject all the incorrect scientific theories. This construction is possible only within the landscape of pure philosophy that furnishes a new Weltanschauung which rejects all the old types of philosophy and goes beyond any particular scientific framework.

In the first part, composed of three chapters, I will introduce the EDWs perspective. Chapter 1 discusses the non-living entities (objects) that belong to EDWs, Chapter 2 deals with living beings (each living being is an EW) and Chapter 3 provides a general view of this perspective.

12 I reject the idea that the job of a philosopher is to raise questions or to make inquiries about certain scientific notions (a rule imposed by Socrates two millenniums ago or by analytic philosophy in the last century). Philosophy simply has to guide science; otherwise, it is not philosophy!

In the second part, I will illustrate the relationships between the EDWs perspective and several specific branches of philosophy and science. In chapter 4, I will investigate various topics belonging to the philosophy of the mind (Descartes' dualism, levels, reductionism and emergence, qualia, Kant and the I). In Chapter 5, I will analyze some approaches and topics belonging to a very recently developed branch of science, cognitive science, and a sub-domain that belongs of this field, cognitive neuroscience. In the first part of this chapter I will demonstrate that neither of the two actual main approaches of cognitive science, computationalism and dynamical systems, do not offer an alternative to the mind-brain problem; here I will also investigate some of the domain's topics/notions in very broad terms (declarative-procedural, accessible-inaccessible, conscious-unconscious, conceptual-sensorimotor, symbolic-subsymbolic, and explicit-implicit, levels, representations, bidirectionality, threshold etc.). In the second part of this chapter, I will illustrate some of the main topics of cognitive neuroscience (CNS, a very dynamic sub-domain in which many people have been working in the last 25 years). The conclusion of Chapters 4 and 5 will illustrate the fact that the mind-brain problem is a pseudo-problem.[13] In Chapter 6, I will present a bird's eye view of cognitive neuroscience and then I will investigate, from the EDWs perspective, some of the main topics of this domain: optimism and its amazing results (Gallant's laboratory) and skepticism (Uttal, Raichle's 'default network'), the localization problem, the binding problems (the main approach, oscillations or synchrony theory), multi-sensory integration, perception and object recognition. In Chapter 7, I will show that the main topic of biology (still unexplained), life, is quite similar to cognition. In Chapter 8, I will deal with Einstein's theory of special and general relativity, quantum mechanics and the relationship between these two theories. My point of view is that these two main theories developed in the area of physics explain EDWs and the 'theory of everything', the superstring theory is just an abstract game realized by physicists who know a lot of pure mathematics with no real application to EDWs.

In the conclusion of the book, after analyzing the main issues of these branches of science ('special sciences', in Fodor's terms), I will explain why I consider that scientists from cognitive (neuro)science, biology and physics need to change the framework within which they work and replace the world/universe with EDWs.[14] By using the EDWs perspective, I will show that many scientific and

13 In my last two books (2012, 2014), after analyzing a vast amount of papers published in the last 4-5 years, I argue that cognitive neuroscience is a pseudo-science.

14 Finally, I will mention that the present book is a synthesis of the main ideas I propounded in my articles (2002, 2005, etc.) and in my books (2008, 2010, 2011, 2012, 2014). I will add that, between 2011 and 2014, several authors published some ideas that are UNBELIEVABLY similar to my ideas, which I published between 2002–2008: Markus Gabriel (2013), whose ideas resemble my entire approach; Georg Northoff (Canada, wrote two books in 2011 and 2014, as well as some papers),

philosophical problems are pseudo-problems created by human beings within the wrong framework, that of the world/universe.

who strangely shares my ideas of the self and cognitive neuroscience; Radu Ioniciou (Romania) and Daniel Terno (Australia) (2011), whose ideas on quantum mechanics strongly resemble my own. I must mention that I posted all my books and many of my articles (all written in English) on several places on the Internet immediately after being published. (For more on these surprisingly similar ideas, see my webpage)

Part I
Epistemologically Different Worlds

1
Principles Concerning the Existence and the Interactions of Objects

In this chapter, I will present and analyze the principles referring to the existence of objects and their interactions, answering questions such as: who determines their existence, where they are, what traits they have and what the relationships between them are, which objects exist and which objects we believe exist, etc. These principles are valid for any set of non-living objects (natural and artificial, or man-made). As I have written in the preceding section, t.he physical (non-living) objects (processes) are not, as it has been assumed until now, in the same world, namely the unicorn world, but they are in EDWs (epistemologically different worlds).

Let us see how these sets of objects and therefore these EDWs appeared. According to the actual physical theories that explain the universe (the unicorn world), after the Big Bang there was the quantum plasma (made of quarks and gluons), which had an extremely high temperature. As the plasma became less and less hot, the first microparticles (photons) escaped from that plasma. Later, the planets appeared in the Universe and much later, life emerged[1] on the surface of at least one planet, the Earth.

This view is constructed within the paradigm of Universe; however, as we will see in the entire book, the notion of Universe/world is completely wrong. Let us see how these sets of objects and therefore these EDWs appeared.

I will introduce the five principles concerning physical objects[2] and their interactions.

(1) Epistemologically different interactions constitute epistemologically different objects, and epistemologically different objects determine epistemologically different interactions.

(2) Any object exists only at 'the surface', due to the interactions that constitute it.

1 We will discuss living organisms, life and cognition in Chapter 7.

2 These principles hold true for any type of object, process or even organism. As we will see in the next chapter, organisms are entities which exist in the macro-EW, where they are a set of physical macro-objects. But their minds or lives belong to different EDWs than this macro-EW where organisms reside. (See the next Chapter). That is why, in the last 3 books, instead of objects I used a more abstract notion, that of entity, which contains all the existent types of objects.

(3) Any object exists in a single EW and interacts only with the objects from the same EW.
(4) Any EW (a set of objects and their interactions) appears from and disappears into nothing.
(5) Any EW is, therefore all EDWs share the same objective reality, even if one EW does not exist for any other EDW.

The existence of a (physical) object generally requires a spatio-temporal framework. Every object exists in one single epistemological world (EW), which means that the object exists and interacts only with objects from the same EW. These notions, existence and interaction/perception, are strongly interrelated. The great English philosopher Berkeley said that „to exist means to be perceived". From my perspective, interaction is a sort of perception, so these two notions are equivalent. So, proposition (1) or Berkeley's slogan can be re-written in the following way: 'To exist means to interact'. Planets existed before man appeared on earth and they will exist even if human beings disappear entirely. Planets (like all macroscopic objects) exist for one another in the macro-EW. This statement is valid for microparticles existing in the micro-EW, as well.

Man is not the only entity who perceives, or who interacts with different objects. If an object is constituted by certain interactions with other objects, what does constitution mean? Interactions constitute the surface of an object. When man sees an object with the help of his eyes, he only actually sees the surface of the object. For example, a man looks at an apple on a table in front of him. He simply sees the apple peel (the apple as a whole), but he does not see anything inside the apple. In order to see what lies inside, the apple needs to be cut. If the man cuts the apple, that apple no longer exists as an object, only two parts of an apple exist.

Now I will make a very important observation: the apple is perceived not just by men, but also by other animals; also, the apple interacts with other objects. Let us suppose that the apple is on a plate placed on a table. As I have written above, we know that the man interacts with (perceives) the plate which, in its turn, interacts with the table. In the EDWs perspective, because the apple, the plate and the table interact (they perceive each other), these objects are in the same EW. Of course, an apple does not interact only with the plate and the table, but it can interact with other objects as well (e.g. with other apples in a fruit basket). The essential thing is that these actions are precisely the ones that constitute the apple, the plate and the table; in other words, these interactions offer objects an ontological status. Without them, the apple (like all objects) would simply not exist. Instead, what would exist would be the microparticles corresponding to the apple, which would interact with each other.

We can use the same reasoning in the case of planets. If there were a single planet in this universe, without anything existing outside of it, that planet would not exist because it would not interact with anything. A planet exists only because it interacts with other planets, in other words, those interactions constitute that planet. It is absurd to claim that the planet would exist 'in itself' or that it would exist for God. Instead, what would exist would be the microparticles corresponding to the planet, since they would interact with each other.

Another question is: how did natural objects, such as planets, appear? According to current physical theories, after the Big Bang the first things that appeared in the universe were microparticles, and planets were formed when a huge amalgam of microparticles were unified. Therefore, can we say that microparticles form a planet? As we have shown until now, the planet does not exist for the microparticles and the microparticles do not exist for the planet, either. Moreover, one of the elementary rules says that two objects (or sets of objects) cannot exist in the same place at the same time. The apple exists only for other apples, for the plate or for the table. The microparticles in it exist, too, but only for other microparticles, not for planets or tables. So there is no point in claiming that microparticles form or compose a table or a planet. Composition, emergence, supervenience and identity are wrong notions that created many other pseudo-notions in various branches of science (for instance, cognitive (neuro)science, physics, biology) and philosophy. Such notions are simply the inventions of the human mind.

That is why we can say that a planet appeared spontaneously out of nothing. The planet's EW appeared out of nothing, but it corresponds to the EW of microparticles. Of course, without the existence of microparticles we would be unable to speak of the existence of planets, but that does not mean that microparticles exist for macroparticles. The macro-EW does not exist for the micro-EW and only man, changing his observation conditions, can observe (indirectly, through correspondence) an EW or another, but these EDWs do not exist for one another. On the contrary, for microparticles, planets don't exist, while for macroparticles it is microparticles which don't exist. Moreover, because of its interactions, only the surface of an object exists, therefore notions like 'internal existence', 'internal determinations', essence are meaningless when it comes to characterizing an object. An object exists only as a whole, i.e., the surface has no parts.

I will offer another example: we are faced with a table. The components of that table (for example, its legs) are not separate from its surface, so they do not exist independently of it. In other words, the legs of a table do not exist as objects. They exist only as parts of the table in the mind of the person who perceives the table at some point, but they do not have any ontological status different from that of the table. If we take the legs of a table away from the table top, the table would cease to exist, but the legs and the top would exist in the same EW as the table, namely the macroscopic EW. (Fig. 4) In other words, the whole does not exist for the parts, nor vice-versa.

Every object has certain traits, characteristics; some characteristics can be perceived by men, others cannot. Moreover, human eyesight assigns to objects certain characteristics which do not actually exist. As we well know, colors do not exist in the objects themselves; color is a perception of the light received by the human eye which is reflected with a certain frequency from the surface of the object. That is why man does not perceive the thing-in-itself (which does not even exist), but, in this case, he has a mental representation of the planets existing in the macro-EW. A planet can perceive/interact with another planet even though we cannot say that a planet observes the same characteristics that a man does. Still, some traits remain the same (what the English 17 c. philosopher Locke called 'first-order' traits), other traits are different ('second-order' traits).[3] Moreover, a bat perceives objects from the macro-EW as having very different traits from those we perceive. For example, colors do not exist for bats. And yet the walls of the cave, for example, exist both for bats and for humans, even if the second-order traits differ greatly. Because EDWs exist or, more precisely, they are, the question 'Which world truly exists?' makes no sense, because all EDWs share the same objective reality.

As we saw in the introduction, one of the main problems in the history of human thought was the relationship between entities. Causality is one of these problematic relationships. Obviously, the notion of relationships is strongly related to the notion of levels. Used under an ontological framework, levels entail causalities which really exist. Used under an epistemological framework, the notion of levels becomes an empty notion, since such levels cannot exist in the same EW. For instance, during the last centuries, there have been strong debates regarding different pairs of levels: the mental level and the neuronal level, (i.e., the mind-brain problem), the micro-level (with microparticles like electrons and protons) or the macro-level (with macro-objects like planets or tables).

If we accept that in such cases both levels exist, we appear to be faced with an ontological contradiction: two objects can exist in the same place, at the same time. Therefore, it is not possible for a table and its microparticles to exist in the same place, at the same time. The acceptance of different types of levels when it comes to notions has created incredible Ptolemaic epicycles (wrong notions and wrong arguments) in the history of human thought. For instance, the notion of 'levels of analysis', used by many actual philosophers, was just a linguistic game which dominated analytical philosophy; the notion of 'levels of organization', used by some scientists and philosophers, led to contradictions regarding the identity of certain entities; and the notion of 'ontological levels', introduced by Descartes, but still used today, produces ontological contradictions within the unicorn world. Therefore, we

3 However, even the distinction between first-order and second-order characteristics is quite artificial.

have to replace levels with EDWs: both such levels exist, but one level does not exist where the other level is concerned, since each level is an EW.

I will draw your attention again to the fact that if we reject the EDWs perspective, contradictions and anomalies will continue to dominate philosophy and science. The scientific or philosophical explanations of some causalities seem to be correct. However, other causalities investigated by scientists and philosophers have produced strong anomalies that created Ptolemaic epicycles. Such Ptolemaic epicycles were formed over millenniums, when a scientist or philosopher thought (incorrectly) that there were certain causalities between objects placed (according to that man's ideas) in the same world. Human beings thought that certain objects were placed in the same spatio-temporal framework when, actually, some of them didn't even exist for others.[4]

Let me return to the quite problematic distinction between the parts and the whole by analyzing some examples. Surprisingly, perceiving for example two objects which appeared to be different, men think that those objects are placed in the same spatial-temporal framework (the unicorn world) and thus seek the relationship (causality) between them. However, those objects do not even exist for each other, so there cannot be any relationship between them, not even one of identity.[5] Obviously, since the causalities are between entities belonging to EDWs, they cannot be explained through generally accepted scientific theories. Again, such anomalies were created because of the single-viewer perspective of human beings: one observer, one world. Within the EDWs perspective, when we try to grasp the relationship between entities belonging to EDWs, in some cases, we have to replace causality, identity and other linguistic notions with correspondences. Below, I will provide some examples.

(1) The example concerning the microparticles and a table (or a planet). As we know, the table (or the planet) and the microparticles exist in EDWs, but the table does not exist for the microparticles, nor the other way around. However, with the EDWs framework, we can say that the table corresponds to that set of microparticles. As we have seen above, we cannot claim that the microparticles form/

4 An extremely important case of this sort is the relationship between waves and microparticles (See Chapter 8 of this book, Vacariu 2008, Vacariu and Vacariu 2010)

5 In other words, it is incorrect to believe that the table is identical with the conglomerate of microparticles, since the table does not exist for the microparticles. Even the notion of identity produces huge problems, especially in cognitive science, when we believe that the mind is identical to the brain. If we accept the theory of identity (the mind is identical to the brain), then is a mental state identical to a certain neuron pattern activating at a certain point of time? This hypothesis is no longer accepted by many people who work in cognitive neuroscience. That is why even the theory of identity was challenged, despite being one of the most accepted theories at this time. Some philosophers (Searle 1992, for instance) proposed alternatives to the theory of identity. (See Chapter 4 of this book)

compose the table because the table does not exist for the microparticles, nor the other way around, so the notion of forming/composing makes no sense. I would like to emphasize that the identity of an object is given neither by its essence or by what it has inside (its composition or other metaphysical, empty notions), nor by the perceptual-constitutive mechanisms of human beings (as Kant and some people working in quantum mechanics believed).

Imagine that someone sends an electron towards a table. (Fig. 1) The question is: what does the electron perceive? A microparticle (the electron for instance) does not perceive, but to perceive is equivalent with to interact. I will ask the reader to imagine that she is the electron sent towards the table. So what does the electron interact with? Most people would answer that it interacts with the table. But this answer is completely wrong because it represents man's point of view. The correct answer relies on the point of view of the electron, not on that of any human observer: the electron interacts with/perceives a huge conglomerate of microparticles which to a human observer represent the table. (Fig. 2)

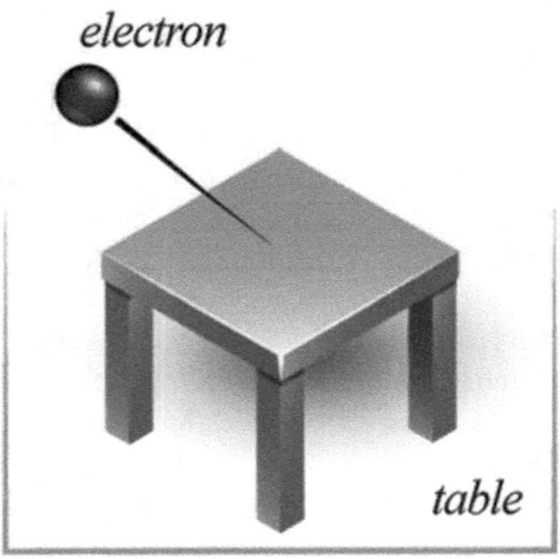

Fig. 1 The wrong image

We believe that a microparticle (an electron, for instance) moves toward the table. We can see the electron only by using an electronic microscope. With the electronic microscope, we do not see the table, but an amalgam of microparticles. For the electron, the table does not exist; only the amalgam of microparticles exists.

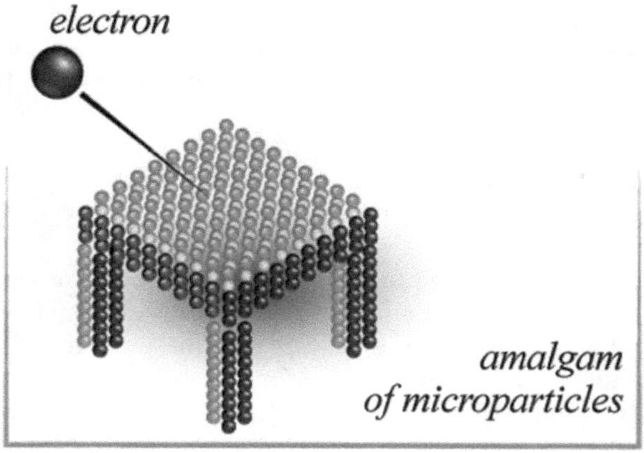

Fig. 2 The correct image

The microparticle moves towards (interacts with) an amalgam of microparticles. (The table does not exist for the electron, but only for our bodies, for other tables, for chairs, stones and other macro-entities)

If we replace the table with a planet we get the same question and the same answer: the planet, and therefore its gravity, does not exist for an electron (there is no quantum gravity because no graviton has been discovered and I believe gravitons do not exist); it perceives only a huge conglomerate of microparticles which, to humans, represents a planet. Moreover, I insist on emphasizing that a planet does not exist for an electron no matter how long the electron travels through the whole universe. In the entire universe, the electron (and any other microparticle) encounters only huge conglomerates of microparticles which men regard as planets. In fact, a planet exists for other planets from the same macro-EW. The micro-particles and the macroparticles (among which planets) are objects/entities belonging to EDWs.

We do not have to break the Kant-Carnap rule using empty notions like gravitons. Those four physical forces (the gravitational force, the strong and weak forces and electromagnetism, which are, in my perspective, equivalent with interactions) belong to EDWs, and the trend to unify them is a huge error which has dominated physics for the past 6-7 decades. A planet simply corresponds to a huge amalgam of microparticles. The understanding of the correspondence between entities that belong to EDWs will lead us to reject the much-desired identity between the mind and the brain or between microentities and macroentities, and the very mysterious (unexplained) relationships between waves and particles in quantum mechanics. Consequently, other essential notions from the various fields of human knowledge have to be rejected: emergence (of all kinds), non-locality and non-spatiality (from quantum mechanics), the relationship between Einstein's theory of relativity and quantum mechanics, etc. It seems that it was easier for philosophers and even for some scientists to play with certain concepts (let us consider merely the Ptolemaic epicycles, or linguistic games – which are often 'the only game in town') in an ideal (or better, ultimate) framework than to recognize the imperfection of a theory. [6]

The question is whether we can explain the gravity of a planet through the properties of certain microparticles (quantum gravity?). For Einstein, gravity is the deformation of space and time around macroscopic entities (planets). Nevertheless, almost all physicists extended gravity from the macroscopic level to the microscopic level and thus invented quantum gravity. No scientist has yet discovered gravitons, but the hope that they will never dies. From the point of view of EDWs, the notion of gravitons is meaningless. (For details, see Vacariu 2008, Vacariu and Vacariu 2010) Within the unicorn world, it seems normal to think that gravity exists at the microscopic level. Nonetheless, quantum gravity is just an illicit extension of macro-gravity and thus it is nothing more than a violation of the Kant-Carnap rule.

Maybe we can say that quantum gravity is a real phenomenon, but due to the (ontological and epistemological) limits of human beings, this notion is impossible to use. As we remember, within the EDWs perspective, epistemology is equivalent to ontology[7], therefore the notion of 'impossible to use' becomes 'impossible to exist'. So, from the perspective of EDWs, we are obliged to extend an epistemological rule (the Kant-Carnap rule) into an ontological principle. To overcome linguistic limitations, we have to realize that quantum gravity really does not exist.[8] Without this rule (proposition), certain amazing Ptolemaic epicycles would continue to dominate science and philosophy.[9]

6 We have to remember Newton, who recognized his inability to explain the strange rotations of the planet Mercury (its epicycles) (Mercury is the closest planet to the Sun). However, he was convinced that somebody, using his gravitational theory, would find such an explanation. In fact, Mercury's epicycle was strongly related to the nature of gravity (another notion which Newton realized he could not explain). More than 200 years later, Einstein demonstrated that the enormous mass of the Sun produces deformations in the spatio-temporal framework in which the planet moves. These strange movements of Mercury could not be explained by Newton's theory of gravity. (For more details, see Chapter 8 or Vacariu 2014)

7 We have to be aware that nature (i.e., the EDWs) has no idea about the distinction between epistemology and ontology made by the human mind. This distinction is the main error which led to the wrong paradigm, the unicorn-world.

8 This step from epistemology to ontology seems to be similar to the movement imposed by famous thinkers (Born, Heisenberg, Dirac, etc.) who created the erroneous bases of quantum mechanics through their explanations of entanglement, non-locality, the probability of reality, etc. The

(2) An automobile and its components (or a table and its components: the table top and the legs – Fig. 4). I have to specify that the wrong notion of composition refers not only to natural entities, but also to human artifacts, like cars. In general, when we speak of a car, we do not refer only to its surface, but also to its internal components. Nevertheless, the car and its components cannot exist in the same spatio-temporal framework, at the same time. Otherwise, we reach an ontological contradiction: two objects would ontologically both exist in the same place, at the same time. More precisely, a car does not exist for its components, nor do the components exist for the car. Still, we cannot claim that the car and its macroscopic components (the engine, doors, windows etc.) ontologically exist in the same EW at the same time. If we did, we would break what I call the ontological rule of parts vs. whole: 'The parts do not exist for the whole, the whole does not exist for the parts'. The car as a whole does not exist for the parts, nor the other way around. When the table as a whole exists (in the macro-EW), its parts (the top, the legs) do not exist. When the table is taken apart, it does not exist at all, only the top and the legs exist in the same macro-EW. (Fig. 4) We will remember that this rule is constructed based on the principles above: the entities are constituted by the interactions that take place.

To us as observers, the car and its parts seem to be the same thing, but the car and its components cannot exist in an ontologically different way at the same time, in the same place. If we accept that both the car and its components exist simultaneously, then we simply reach an ontological contradiction: two objects cannot both exist in the same place at the same time. Therefore, ontologically speaking, in one EW at one moment in time you can either have the car, or its components, but not both. Essentially, we, human beings, have no right to decide what exists and what does not. Existence is constituted by the interaction each entity has with the other entities in the same EW, which is naturally true of the car and its components. Since both the car and the components belong to the same EW, they cannot exist at the same time.[10]

difference is that the last step within the unicorn world produces great anomalies (see the last chapter in Vacariu 2008, and Vacariu and Vacariu 2010).

9 „Imagination is more important than knowledge.“ (Einstein) Nevertheless, we really need to impose certain hyperontological constrains on the human imagination. Otherwise, we misplace the real knowledge we have about reality in a surrealist surrounding.

10 Some might consider that the car (or its components) exists only because of the functions it performs. This would reduce ontology to functionalism, which would be completely wrong. The EDWs perspective refers to that which exists without the presence of the knowing human subject, namely to the ontological status of entities (which exists in the absence of humans). Therefore, functionalism (an approach which requires the presence of the human subject) is completely averted, as are other theories related to this research area.

The same is true for the relationship between the car and the microparticles[11], which are also components, but microscopical ones: if we consider that both exist (in the same EW), we break the ontological rule of parts vs. whole. However, in this case, the whole (the car or the table) and the parts (the microparticles and their micro-forces) are in EDWs, not in the same EW as the car and its macroscopic components. For the relationships between a macroscopic object and a set of microparticles (and also for the parts-whole relationship), we have to replace the notion of composition with that of correspondence: an automobile is not composed of its parts because the automobile does not exist for its parts (and vice-versa). A table is not composed of microparticles because it only corresponds (ontologically) to a set of microparticles (and their micro-forces) since the table (and other macroobjects and their force, gravity) and the microparticles (and their micro-forces) belong to EDWs. EDWs in images:

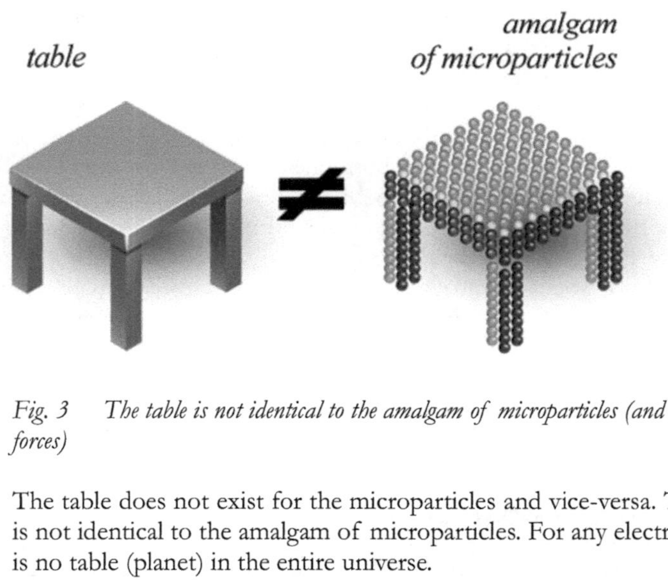

Fig. 3 The table is not identical to the amalgam of microparticles (and their micro-forces)

The table does not exist for the microparticles and vice-versa. The table is not identical to the amalgam of microparticles. For any electron, there is no table (planet) in the entire universe.

This distinction between the parts and the whole is a philosophical distinction which created many metaphysical (linguistic) games that have nothing in common with nature. What really exist in EDWs and what we believe that exist are very

11 The same rule applies to any macroscopic object (such as the table) and to microparticles.

different things. Again, human language/thought had a dictatorial status in establishing the dominance of the unicorn world. The conclusion of investigating all these examples is that the parts and the whole exist (a) in some cases in EDWs: for instance, the microparticles and the macroparticles (b) in other cases, in the same EW but not at the same time: the whole does not exist for its components and vice-versa. For instance, if a table exists with its components (its legs and its top) at the same time, there would be an ontological contradiction: two (sets of) entities exist in the same place, at the same time. The whole vs. parts relationship in images:

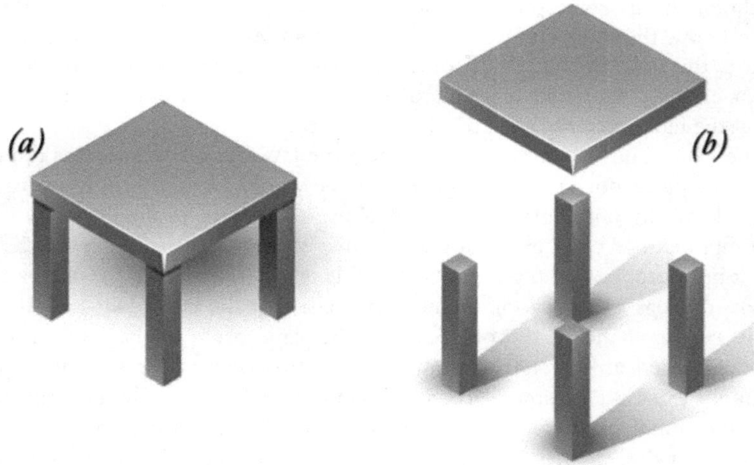

Fig. 4 The whole vs. parts relationship (an admittedly incorrect expression, since such a relationship does not really exist)

(a) Only the table exists in the macro-EW. As ontological entities, the top and the legs do not exist.
(b) A top and four legs exist in the same macro-EW. As an ontological entity, the table does not exist.

In principle (5), when I claimed that 'all EDWs have the same objective reality', I used the Kantian expression of 'objective reality' with almost the same meaning that Kant has given it. The 'conditions of observation' are replaced by 'conditions of interaction' and thus the Kantian epistemological notion of constitution (even if 'ontologically loaded') becomes a true ontological-epistemological concept available to EDWs. The first part of principle (5) is 'any EW is'. An EW does not exist (i.e. it does not exist in a spatio-temporal framework), but it is. Only certain epistemologically different objects (and their interactions, but an EW does not interact) exist within their spatio-temporal framework. So, an EW is, an object exists (in a spatial-temporal framework). Using different conditions of observation (our eyes or an electronic microscope, for instance), we see that macroparticles and microparticles exist in the same spatio-temporal framework, but in EDWs. All the epistemologically different entities and their epistemologically different interactions, that is all EDWs, have the same objective reality.

Let us summarize the parts-whole relationship:

(1) The whole and the parts can belong to EDWs. For example, a table (the whole) and its corresponding microparticles (the parts) belong to EDWs. As we know, the interactions between entities are essential to each EW. In this case, the table interacts with other macro-objects, but the microparticles do not exist for the table. The microparticles interact with other microparticles, but the table does not exist for the microparticles.

(2) The whole and the parts belong to the same EW, but the whole does not exist for the parts, nor do the parts exist for the whole. More precisely, we cannot speak of the ontology of the table at the same time as the ontology of its components. At any given moment, either the table exists, or its parts exist, but not both. For example, either the table exists (the whole), but the parts do not (the legs and the top), or the parts exist, but the table does not. If both the table and the parts would exist in the same place, at the same time, we would reach an ontological contradiction: two objects (or sets of objects) cannot exist in the same place, at the same time. Let us apply this framework to the situation of cars and their components. We see a car going down the street. At that moment we cannot claim to see certain parts going down the street. (I am not referring here to functionalism.) For us, the car exists as a whole. The parts cannot exist, because we would reach an ontological contradiction. If we see that one of the rear-view mirrors is broken, we can say that a part of that whole does not exist, but we do not refer to a component of the car. The car and its parts exist in the same EW, but not at the same time. If we discussed the components and the whole, we would reach, again, an ontological contradiction. The same rule applies to the brain as a whole and its components.

(3) The mind and the brain are or exist in EDWs. In other words, the mind is an EW, the brain belongs to the macro-EW. (see the next chapter)

2
Principles Concerning the Brain/Body and the Corresponding I (the Self/Mind)

In this chapter, I will explain the principles referring to the body (to the organism in general, brain included) and the corresponding I (i.e., the self or mind). Generally, any body/organism exists in the macro-EW along with other macroparticles. For instance, our bodies exist in the same macro-EW as tables and planets. As we saw in the previous chapter, microparticles do not compose our bodies, but they correspond to them, since they belong to the micro-EW. What are the relationships between a body and its cells, between an organism and its life, between a cell and its life, between a human body and the mind? As we will see in chapter 3, it has been wrongly claimed that the organism has life, and since biologists consider that the smallest living entity is the cell, therefore a cell has life, too.

There are people who consider that there are at least strong similarities, if not an identity, between life and mind. (See the next chapter of this book) Therefore, seeing the situation from that perspective, the statement 'An organism has life' is quite similar to 'The brain has mind'.[1] In biology, after 100 years of research, nobody can describe the true relationship between an organism (or a cell) and life. Cognitive neuroscience has been unable to solve this problem after four decades of work, despite the efforts of many scientists (neuroscientists, psychologists, computer scientists, etc.) using various apparatuses for scanning or reading the brain. Therefore, ontologically speaking, within the EDWs perspective, there are no great differences between life and the mind: both are 'living beings'.

As we have already seen in the previous chapter, human entities are not the only observers within the EDW framework. Each entity has its own 'point of view', or, better said, it 'interacts with' other entities belonging to the same EW. Even in the unicorn world, certain animals have points of view that are more or less different from the human one. More precisely, many animals (e.g., the bat) perceive the external world quite differently from human organisms. From animals, we can

1 If we accept the claims made by Searle (1992) that „The brain produces the mind", we come across great anomalies. This statement has led to great debates in philosophy and cognitive neuroscience. It is not surprising that, working within the unicorn world, philosophers have been unable to solve the mind-body (brain) problem in over 300 years!

move on to less and less complex living beings, such as bacteria, both multicellular and unicellular, whose perspective differs even more strongly from the human one.

If we accept the postulate that the smallest entity that has life (that life emerges from) is the cell, we can wonder how a cell perceives its surrounding environment. While some may find this question ridiculous, we have to remember that, within the EDWs, even an electron or a planet perceives (i. e., interacts with) other entities from the same EW. So, if we consider that a bat perceives its environment, then we can also consider that a unicellular organism perceives its surroundings or that a cell interacts with/perceives its external environment. However, in this context, the question is 'What is the environment of life?' (I am not referring to the organism). As we will see below, the concept of 'living being' is quite similar or even identical to that of the mind, so life is an EW. In fact, I consider that life and the mind are identical. The principles regarding the I/self and the body (organism/cell) are the following:

(6) The I/self/mind (life) corresponds to a body (organism/cell).[2] The self does not exist for the body, the body does not exist for the self.
(7) The I is an EW. Therefore, the I is in a temporal framework, while the body exists (in a spatio-temporal framework).
(8) Having a certain set of components, from our point of view the body corresponds to (but it is not composed of) an amalgam of macro-objects (arms, legs etc.) (or cells) and their relationships. The body and its corresponding parts (or cells) belong to the same macro-EW. Also, a body corresponds to a certain set of microparticles belonging to the micro-EW.
(9) Certain mental states and processes form the knowledge that is the I.
(10) As an entity, the I has unity as an indeterminate individuality (it does not have spatial dimensions).

The human body is neither more, nor less than an entity like all other entities: the body is something epistemologically different from the self (the mind or the life): the body exists in the macro-EW, while the self (the I or the mind) is an EW. Various instruments of observation (or measurement) simply allow the I to perceive EDWs indirectly (i.e., through correspondences). However, there are no relationships between the body and its micro-entities (or its cells): each micro-particle interacts with an amalgam of other microparticles (which belong to the micro-EW) and they all correspond to our body (which belongs to the macro-EW). Moreover, the body/brain (which belongs to the macro-EW, where we can find planets, tables, and other macroparticles) does not exist for the self (the I) which is

2 These principles are also valid for life-organism/cell. (See Chapter 7 of this book)

an EW. The self (the living being) requires the existence of the corresponding brain (body). Without this correspondence, the I/mind cannot be.

The I (the self) is an entity and an EW at the same time. This is the only case in which we encounter the paradoxical situation in which something is both an entity and an EW. Obviously, since the self (the mind) corresponds to an organism/cell and the self is an EW, then like any EW, the self appears from and disappears into nothing (or it comes from nowhere and returns there).

One of the essential features of the correspondence between the I and the body/brain is that, in its absence, the body (which includes the brain) would be unable to survive in its environment. We cannot continue to maintain, as we have done thus far, that the biological functions (which are the results of evolution) are enough for an organism (a body) to survive in its environment. An organism has no knowledge about its external environment, even if the body strongly interacts with entities that belong to that environment. Such interactions do not represent knowledge: a planet interacts with another planet, but could we consider that planets have/possess knowledge because of that? Also, the body (brain) has no knowledge at all. Does a neuron perceive color or space? Does a neuron feel time? The neurons (and the white matter and all other components of the brain) do not posses any kind of knowledge at all. There are only electrical and chemical (and other kinds of) reactions, but not knowledge. Certain physical (neuronal) processes take place, but that is all.

Within the unicorn world, the coordination of all the biological functions of an organism is thought to be the result of evolution. The evolution of an organism is explained through this coordination. From the EDWs perspective, the coordination of all biological functions needs a unity that is impossible to use within the biological mechanisms of an organism. Only the self (the I or the mind) has this kind of unity, which corresponds to some biological mechanisms, but which does not exist within an organism.[3] Even if there is a correspondence between this unity and the neural and bodily mechanisms/functions of an organism, this unity is neither more, nor less than the I/self. Moreover, this unity corresponds to the development of an organism and the evolution of a species in a particular environment. We return again to our analogy: can we find the unity of a table (i.e., for the standard table, its rectangular surface placed on four legs) within its microparticles? Obviously, there is no such surface (with certain features like area, color, etc.) within the EW of microparticles.

I believe that the scientists working in the field of cognitive neuroscience are making a major mistake when they avoid taking into account the essential role of

3 This is the reason why the binding problem or localization (and many other problems of cognitive neuroscience) are pseudo-problems. (For details, see Chapter 6 of this book or, for more details, see Vacariu 2012)

development and evolution when they analyze the relationship between the mind and the brain/body (especially using brain imaging with fMRI and PET). For instance, when we use fMRI for brain imaging (for scanning or reading the brain), we cannot grasp the corresponding results in the development of an organism and in the evolution of the species. These corresponding results would represent the I/self/living being.

Many mental and behavioral functions appear during the development of the corresponding organism. After a period of training (which can last weeks, months or years), many neural areas get activated less when they perform certain functions. In another EW, the mind-EW, in psychological terms, the explicit knowledge (the conscious or declarative knowledge) is transformed into the implicit (unconscious or procedural) knowledge in order to perform certain tasks (such as speaking a foreign language, riding a bicycle, etc.). After the training, these tasks are accomplished better, so more training leads to better accomplishments.

The problem is that when we scan the brain of an adult, we cannot observe the activation of all these neural processes, of chemical and electrical reactions, of neural waves, etc. However, the required mental functions are performed without any problem. Certain neuronal patterns (which were the most strongly activated at the beginning of that task) seem to be inactive, or at least they are much less active. Most probably, because of their habitation (training), the neural patterns are less activated, but not completely inactive, since the task is accomplished more efficiently. The same processes appear during the evolution of the species. Therefore, in order to achieve a task, many parts of the brain are involved[4], but we are unable to observe all of them.

One of our major mistakes is that we correlate some mental functions only with certain neural functions which we observe using fMRI and PET. In reality, if we go back in time (to the development of an organism and the evolution of the species) we will notice that many other neural areas are involved in performing a certain mental function—in fact, the whole brain and body participate. In these conditions, it is futile to try to identify what oscillations, chemical reactions and many other processes (not only activations of neuronal patterns, which can be seen using fMRI) take place inside the brain in order to correlate any mental task.

Why do we think that when we use fMRI we can actually read the brain? We read only the top of the brain processes, we introduce some labels (i.e., we correlate the activations of some neuronal patterns with some mental tasks that belong to the

4 Do not forget, the brain does not have any relationship with its parts! I try to use approximately the same language that I used until I discovered the EDWs. That is, I have tried to introduce as few new concepts as possible. Even if a linguistic change would be necessary, if I were to break away entirely from the old, common language, nobody would understand anything I wrote in the five books I have already published.

I) and then we claim that we read the brain. For instance, if we could scan the brain of a child during its first months of life, we would probably observe the activation of numerous parts of the brain even for the simplest movements of the arms, the legs, or even the eyes. After a period of training, many parts of the brain will cease to be activated as strongly by such actions. Nevertheless, those tasks are performed better and better. Consequently, using fMRI and PET, we can find only certain neural areas, maybe the most activated ones, but not all those which are correlated with certain mental functions.[5] Moreover, we cannot observe the activations of some neurons and their oscillations with the same apparatus.[6]

Each mental function (each mental representation) belongs to the unity of the I/self. This unity represents the indeterminate individuality of the I/self, or better said, the I is an indeterminate individuality. This means that the I/self/mind (which is also life, see the next chapter) has no spatial framework, only a temporal one. In this sense, we cannot identify the life or the mind as being placed within a spatial framework. No matter what conditions of observation we choose, we will not be able to identify the individuality of the life/mind/self.[7] Therefore, this individuality is indeterminate (not non-determinate). Trying to reduce the indeterminate individuality to a completely determinate individuality (which would presuppose at least a spatial or spatio-temporal framework) would mean using a mixture of EDWs. We can talk about color or surface only when we refer to entities placed within such a framework, but can we consider that mental states (for instance, emotions or beliefs) have spatial dimensions?

If we were able to perceive certain determinations (within a spatio-temporal framework) of the I/self, it would mean that we could determine the individuality of self. However, this action would break the Kant-Carnap rule. Moreover, as we will see below, it is impossible to construct instruments to observe/perceive the

5 A quite famous cognitive neuroscientist, Uttal, considers that the entire brain (or large parts of the brain) is activated for any mental task. (Uttal 2011; see Vacariu 2012) I marked Uttal as a pessimist in today's cognitive neuroscience. Uttal replied to me (in our private correspondence) that he is not a pessimist, but a realist!

6 This idea reminds me of Bohr's complementarity principle, created to explain the phenomena described by quantum mechanics: we cannot observe the wave and the particles by using the same apparatus at the same time. In fact, within my EDWs perspective, I directly applied Bohr's principle concerning waves and particles to the mind-brain problem. (See Vacariu 2005 or 2008)

7 This problem is related to the distinction between organisms and life. We cannot identify life in an organism because life (the mind) is an EDW (without spatial dimensions) than the macro-EW (with spatial dimension) in which the organism/body is situated. (For more concerning life and organisms, see chapter 7) We cannot identify the mind inside the brain because the mind is an EDW (without spatial dimension) than the macro-EW (with spatial dimensions) where the brain/body is situated. We cannot even say that the mind is identical to the brain because they are entities which have epistemologically (i.e., ontologically) different properties and they are placed in epistemologically different (spatio-)temporal frameworks.

mind/self (or its unity) as a whole. Once more, within the brain (or the body), we cannot find a unit that corresponds to the self/mind. In reality, the situation is even worse: actually, we cannot identify (or better said, correlate) any mental task with a certain neuronal pattern. Therefore, within the brain (body), the 'indeterminate individuality' is meaningless. Any brain/body (or its parts) has certain spatial (and obviously temporal) determinations, but the self/mind is an indeterminate individuality.

From my point of view, a body or an organism (which corresponds to the self/living being) is composed of cells (each cell also corresponds to a living being and all cells correspond to the body). The organism and the cells all exist in the same EW (usually, the macro-EW), but not at the same time. That is, we cannot break the parts-whole ontological rule: 'The parts do not exist for the whole, the whole does not exist for the parts'. Moreover, the body does not exist for the I/living being (the self/mind), and the I does not exist for any organism. There are some interactions between the brain/body and external objects, but these interactions correspond to certain mental representations that belong to (or better said, they are) the I/self.

The living being (the self) cannot be identified through any kind of perception (or its extension) because all human perceptions are the I/mind, so they belong to EDWs. The fact that all perceptions are the I is required by the unity of the self/mind. If all perceptions were not the I or, in other words, the I would have access to or perceived visual input, for example, then we would lose the unity of the I/self and we would need to introduce the notion of a homunculus (a tiny man situated in the brain), or internal eye. If all entities (except the self/living being, or any EW as a whole) can be perceived, then we might imagine a 6 sense which perceives the I. Can we hope that in the future humans will create special instruments to achieve this perception? If that were the case, the I would be an entity with certain (spatial) determinations. Would this be even theoretically possible? From the EDWs perspective, having a 6th sense which perceives the I/self is an ontological contradiction and, moreover, the self would lose its unity; therefore, the corresponding organism would not be able to survive in its environment. We can conclude that the construction of such an instrument is impossible, because, as I have said above, 'the I is' ('the mind/self is'), or 'the I is an indeterminate individuality' (this can also be stated as 'the mind/self is an indeterminate individuality').

Only within the unicorn-world could we say that a biological organism possesses knowledge. Within the EDWs perspective, however, we have a correspondence between any brain/body (a human biological organism or cell) and the self/living being. However, knowledge has nothing to do with the organism, obviously, but only with the I/self/living being. In this case, it is wrong to use the sentence 'A being has knowledge' for at least two reasons. The first reason is that it is impossible to refer to a being but only to the being/self/mind, since one being/self/mind does not exist for any other being/self/mind (they belong to EDWs). The second reason is that it would imply a difference between the being and its knowledge. This linguistic difference is very wrong, indeed,[8] because it would make us lose the identity of the self/life.

Another observation: there is nothing inside or outside of the self/the I. The proper way of phrasing this situation is 'Knowledge the self'. If we say 'The self possesses knowledge' or 'the self perceives something internal to the body', then the self does not have the unity required (through correspondences) for the development of an individual and for the evolution of the species. Without this unity, the self would not survive during the development of an individual, so the organism would not survive the evolution of the species.

We can find this unity by looking at the self/mind corresponding to the organism/body. Moreover, there are different types of knowledge (declarative and procedural, implicit and explicit, conscious and non-conscious, etc. – see Vacariu 2008), but these types of knowledge do not form/compose the I (i.e., they are not parts of the self). All types of knowledge are the self/mind/life. The entire knowledge of human beings is not of the self/life, but it is the self/life. Otherwise, we would be faced with the great problematic difference between knowledge and the I, namely that the self would be decomposable (it would have no unity), which, as I have said above, would not allow either the proper development of an individual, nor the evolution of the species. Without this unity of the self, we would need to find the spatial dimensions of the self/mind/life, which obviously do not exist. We have to remember the paradoxical status of the I: it is an entity and an EW at the same time.

Here is a thought experiment which will help you understand this through an analogy: that of the subjectivity of a planet. Imagine that you are a planet that is unable to observe itself. Paradoxically, your perception is restricted to microparticles alone. Consequently, you as a macroscopic object cannot observe any macroscopic object. Such entities simply do not exist as far as your body is concerned. This situation is similar to that of the I/self: the I cannot observe/perceive itself. Let us suppose that you are a reductionist and an empiricist, so you think that only microparticles exist. However, there are some phenomena (for instance, those that correspond to the gravitational forces of macroscopic objects such as planets) which cannot be explained through microparticles and their interactions. The gravitational force is related only to macro-objects (like yourself), not to the micro-

8 This distinction leads us to a regression ad infinitum in looking for the homunculus, the 'tiny man' or the internal eye that perceives, inside the brain or the mind, any mental representation corresponding to an external object. I repeat, the 'external object' refers to the brain/body, not the I (which is an EW).

particles, since nobody discovered gravitons (the microparticles which would produce gravity). Without being able to explain gravity (a real phenomenon) by using microparticles and micro-forces, you (the planet) will use 'dark matter' and 'dark energy'.

The I (the living being) is similar to the planet from this thought experiment. We can perceive indirectly, through correspondence, a planet (or a table) with the corresponding eyes that belong to the body situated within the corresponding EW (the macro-EW), but we cannot perceive the self/mind/life because each of them is the self/I. However, a self does not exist (more correctly, it 'is not') for any other self (since the self is an EW) so, linguistically, it is meaningless to talk about a self. It is better to refer to 'the self' and not to 'a self'.[9] All mental perceptions (which correspond to the functions of certain biological mechanisms) are the I. Therefore, the I/self cannot perceive itself. It is like asking an eye to see itself. In such a situation, it is impossible to conceive of a 6 sense which would perceive the being. To do so would lead to a contradiction. I will call this inconsistency the 'being-perceiving' contradiction, which leads to this rule: 'The I/the self/mind/life cannot perceive'. Nobody considers that life as such perceives something, but many people believe that the mind perceives the external world. However, the mind, just like life, does not perceive anything. There are two arguments to support this rule:

(1) The I/self/mind does not perceive itself or any entity from any EW, since the self (the I) would need a biological mechanism to perceive something, which would require a mixture of EDWs (this is, again, an ontological contradiction). A biological mechanism and its activities merely correspond to the mental perceptions that are the I/self (all mental states, including perceptions, are the I). The biological mechanism cannot perceive itself, otherwise we would be faced with the being-perceiving contradiction. Therefore, the I/self is an 'indeterminate individuality' (it has no spatial dimensions). Mental perceptions are the I, but there are no mental representations for perceiving the I. Any mental perceptual state is the I, but it appears spontaneously and it corresponds to neuronal and biological (the eyes and subcortical areas) mechanisms and processes which interact frequently with external objects (all the objects and the bodies/organisms are situated within the same environment, the macro-EW).

9 If we take this rule further, we cannot talk even about EDWs, since one EW does not exist for any other EW! Because of this, I will stop here, or we would have to give up a large part of our common and scientific language, which would make communication impossible. Let us suppose that two people are communicating (speaking) using a particular language. For those two organisms (ears, tongues, eyes, brain, etc.), the 'spoken words' of a language are not 'real words' but just audio and/or visual signals. The words (with their meaning and syntax) exist only for each self. More precisely, the words (e.g. a memory of any sort) are the I (an EW) and correspond to the audio and visual signals produced by an organism and received by the other organism situated within the same macro-EW.

(2) As an EW, the I/self/living being obviously cannot perceive something else. For instance, the mind cannot perceive another mind, because the mind is an EW and interacting with another EW would lead to a mixture of EDWs. The mind cannot perceive any macro-object or micro-object because this would lead to a mixture of EDWs between the mind (an EW) and the macro-EW (where the macro-object is situated). Idioms like 'inside my mind' or 'what is on your mind' are just unregimented linguistic slogans created within the framework of the unicorn world. The interaction of two minds/lives would automatically represent an ontological contradiction. More precisely, a living being cannot observe another living being because every living being is an EW and there are no pluralities of living beings. As I have said before, one living being does not exist for any other living being, exactly as a mind does not exist for any other mind: the mind or life or living being is an EW and one EW does not exist for any other EW. There are organisms (bodies) which can interact within the same macro-EW, but the minds do not even exist for one another, so how could a mind/life interact with another mind/life?

We can analyze another example which further clarifies this contradiction: 'I perceive my hand'. As we already know, the I is an EW, the hand is part of an organism that belongs to the macro-EW. What does 'I perceive my hand' mean? With the help of light, the eyes (which are extensions of the brain) interact with the hand. The eyes, the brain and the hands are all body parts (pay attention to the 'parts-whole' ontological contradiction). The I is an EW, while the body (brain) belongs to the macro-EW. So, the interactions between the external objects, the hand, the light waves, the eyes and the brain/body, all of which belong to the macro-EW, only correspond to the mental perceptual states that appear spontaneously in the mind-EW, but they are the mind-EW (or the self/the I).[10]

This leads us to the conclusion that it is impossible for the self/living being to perceive anything. It is wrong to assert that 'The I perceives a macro-object situated in its surrounding environment'. The I does not perceive anything, since percep-

10 For a better understanding of the EDWs perspective, the reader can recall the analogy between pairs such as microparticles-planet and brain-mind. Almost everybody would agree that microparticles do not produce or cause the existence of a planet, nor does the planet emerge from the microparticles. Moreover, the microparticles are not identical to the planet, since there are different forces acting on the level of microparticles (quantum forces) than on the level of the planet (gravity). The same holds for the other dichotomy, brain-mind: the brain does not produce the mind (Searle 1992); the mind does not emerge from the brain. (concerning emergence, see Chapter 4 of this book or Vacariu 2008) The mind is not identical to the brain, since the two have completely different features: the mind is an EW, the brain belongs to the macro-EW. Following this analogy, the microparticles and the macroparticles belong to EDWs.

tions are the self/living being that is an EW and cannot interact with/perceive objects that belong to another EW.

What happens is that certain perceptual images (perceptual representations) correspond to certain neuronal and bodily processes which appear due to interactions with the external environment, but mental representations are not formed through the interactions between the mind and the external world. Again, perceptions are the I/mind and they correspond to the interactions between the organism (or the body, which includes the brain) and its external environment (the macro-EW where the body is situated). Perceptions are mental states and they are the I, not parts of the I. If the I would observe mental perceptions, it would require a homunculus or internal eye to perceive them, but the homunculus would lead to certain ontological contradictions. I strongly insist that, from an EDWs perspective, the notion of perceptions is incorrect. Nothing is perceived because that would presuppose an entity which perceives, as well as one which is perceived; perception is an EW (the I) and an EW cannot be perceived. The I/mind always corresponds to an organism/body.[11] Only within the EDWs framework can we avoid huge errors in our thinking.

The conclusion is as follows: perception is the I/living being that is both an EW and an indeterminate individuality (mainly, it has no spatial dimensions).[12] Various perceptions (and feelings) are the entities of a living being, which is an EW, but their individuality/identity is epistemologically and ontologically different from the individuality/identity of objects or organisms. The main difference is caused by the status of being, the indeterminate individuality. Every type of perception is the self/living being which is an indeterminate individuality. However, a perception has a kind of individuation different from that of any object or organism. Generally, the individuations of objects or organisms are placed within a spatio-temporal framework, while perceptions are spontaneously different in time, but not in their relationship with the whole self (which has no spatial dimensions, only a temporal one).

11 An alternative is Berkeley's idealism, but accepting it today would require us to turn to the church, not to an academic (philosophical or scientific) institute.

12 In the case of some EDWs (the macro-EW, the micro-EW, etc.), the objects inside the EW have spatial and perhaps temporal dimensions (photons, for example do not have a temporal dimension). However, no EW can have spatial dimensions as a whole, because that would require someone to observe it and this would lead to a mixture of EDWs. Objects are usually placed within a spatio-temporal framework inside an EW, but it is meaningless to talk about the spatio-temporal framework in which the EW itself is situated. Otherwise, the EW would interact with another EW, which is an impossible ontological contradiction. The notions of 'many worlds' and of 'parallel worlds' are simply the inventions of great 20th century physicists! (See Chapter 8 of this book, Vacariu 2008, Vacariu and Vacariu 2010)

These perceptions are not internal perceptions13 of the I/mind; instead, they are the I.

Perceptions correspond to certain neuronal patterns of activation (the most activated ones) and to the patterns of the rest of the brain and the body, but we have to take into account the fact that the brain has the property of superpositional storage (Clark 1993, 1997 or see Vacariu 2008), exactly like a connectionist or neural network: various kinds of information overlap over the same network. If we accept the identity theory, this superpositional storage would not allow us to make the individualization of mental perceptions in our mind. Only the idea that mind and brain/body are an EW or belong to an EDW offers us the possibility to explain the individualization (in a temporal, non-spatial framework) of perceptions, which are the mind. As an EW, certain entities (mental states and processes) are the I/mind. It is completely wrong to consider, as many people do, that the I has certain feelings or perceptions since, from my point of view, visual representations or emotions are the I.

If we accept that biological mechanisms produce (cause) perceptions, we infer that the mind is the product of the brain. In this way, we return to Searle's „rediscovery of the mind" (1992). This rediscovery is actually a complicated Ptolemaic epicycle constructed within the unicorn world. (For details about Searle's philosophy, see Vacariu 2008) Production would require causality between the brain and the mind, which would involve a mixture of EDWs. In this case, as in many others, we have to replace causality with correspondence. Understanding this replacement is a step towards accepting the EDWs perspective. Again, it is meaningless to look for any relationship between the I (mind or life) and the human body/brain, or for any interactions between the I and the external world.

I took the idea to identify all mental perceptions with the self (the I) from Kant's transcendental philosophy.[14] However, the great problem for Kant was the noumenal-phenomenal distinction, a distinction imposed by the unicorn world. By using the EDWs perspective along with Konrad Lorenz's idea (1941) regarding the adaptation of the organism to its external environment during the evolution of the

13 Again, the internal perceptions need the eternal homunculus or internal eye! Interestingly, working within the unicorn world a long time ago, Descartes was forced to introduce this notion. (See Vacariu 2008)

14 „The outcome of Kant's theory of understanding could therefore be expressed as follows: the world is not simply my world, as with other subjective idealist philosophers; the world, for Kant, actually is the self." (Waxman 1995, p. 857 in Vacariu 2008) If the world (i.e., the perceptions of the external world) would be my world (i.e., my perceptual states) inside my mind, an 'internal eye' (a homunculus) would be required to look at a TV screen inside the brain or the mind. However, there is no TV set in the brain (we have no green color in the brain), nor in the mind (we have no spatial dimensions in the mind). The perceptual images of the 'external world' are knowledge, which is the self, which is an EDW than the macro-EW in which the body/brain are situated. (For more details about Kant's transcendental philosophy, see Vacariu 2008)

species, the problematic Kantian noumenal-phenomenal distinction is completely avoided. Moreover, the brain and the body evolve together in a strong relationship (see Sporns 2006) during their interactions with the external environment.

Nonetheless, the mind (the I) corresponds to the body (brain), so the famous Kantian distinction between pure and empirical intuitions of space and time is meaningless, too. The empirical intuition, which is a part of perception, presupposes interactions taking place between the I and the world, which we have already seen do not exist. All elements of perception, including the intuitions of space and time, are the I, while the corresponding biological mechanisms evolved in connection with the macro-EW (an EW that really is). Obviously, within the unicorn world, it was not possible for Kant to construct a better philosophy.[15] Nonetheless, the noumenal is meaningless within the EDWs, exactly as Newton's absolute space and time are meaningless for Einstein's theory of relativity, in which space and time are always relative. (Concerning Einstein's theory of relativity, see Chapter 7 of this book or, for more details, see Vacariu 2014) Moreover, we have to discard the strong differences between the ontological statuses of mental representations and external objects: all entities, whether objects or mental representations, have the same ontological (more correctly, epistemological-ontological) status but all these entities belong to EDWs.

The perceptual representations do not reflect the characteristics of external objects precisely, but nevertheless, the similarity between these perceptual mental states and the external objects in question is quite strong. This similarity mirrors the similarity between a TV screen (2-dimensional coordinates) and the images on the retina (also bidimensional figures), but as we noted above, there are no images on the retina, exactly as there are no real images/photos (a bird flying, for instance) on the TV screen. The screen only has certain activated pixels (not a real bird, nor a representation of that bird). The same situation is true in the case of the retina or the brain: the first merely has certain photoreceptors activated; for the second, certain neuronal patterns are activated (as well as other neuronal and biological mechanisms).

It needs to be clear that there is no representation of that bird anywhere inside the head (body/organism). It only exists in the mind. More precisely, this representation is the self/mind. However, the mind has memory. Can we consider that memory has spatial dimensions? In other words, when we remember the name of a city (for instance Prague), is it spatially placed in our mind? Does the word justice have spatial coordinates? When we remember the image of a dog, is it like a photo in our minds? Does this photo have colors, size and borders? Obviously, the answers to all these questions are negative because the mind-EW (the I) is an EW

15 Kant's approach had a major, long-lasting effect in both science and philosophy. In some ways, it strongly influenced even contemporary philosophy.

that has no spatial dimensions, just a temporal one. We can ask similar questions not only concerning memory, but also concerning other functions of the human mind (perception, language, imagining, abstract thinking, etc.) and the answer will be the same: no mental function is situated in space.

Let us analyze another case in more detail: a human perceives a house. „How do we know where objects are located in the world? When we look at the world, the image that strikes the back of our eye is essentially two-dimensional, similar to the image that would be taken by a camera. This two-dimensional map of the world projected onto the eye is preserved in the early visual areas of the cerebral cortex, which provides a map of where objects are located relative to the center of gaze. The brain is also able to figure out the missing third dimension and estimate how far away objects are in space." (Baars and Gage 2010) The light is reflected by the house, then it reaches the eye. It crosses through the lens of the eye and hits the retina, which leads to the activation of certain photoreceptors (rods and cones responding to different kinds of light). They say that the image of the house is reverted on the retina.

In reality, we cannot talk about real images on the retina. Those images on the retina (which is a part of the eye, which is a part of the body) are active patterns of photoreceptors whose display corresponds to the proportions of the signals produced by the photons (which correspond to the wave-light) reflected by the house, which has a certain size. There is no reversed image of that house on the retina. The activations of photoreceptors are just biological processes and nothing more. Does the brain interpret the activations of some photoreceptors of the retina and some neuronal patterns of cortical areas as being images? This is the wrong question. The bidimensional image on the retina is sent, through the lateral geniculate nucleus (LGN) and through other sub-cortical areas, to the visual cortical area. Can we believe the images from the retina are still preserved in the visual cortical area? There is no image of that house neither on the retina, nor in the LGN, nor in the visual area. These neuronal processes are just neuronal processes, not mental representations.

Therefore, the answer to the question: 'Where can we find the 3D mental representation of that house?' is that we cannot, neither in the eye (it does not exist on the retina), nor anywhere in the brain. This representation, like all other mental visual representations, is the mind/the I. Obviously, certain neural processes and the states of the entire brain and body simply correspond (with a high degree of approximation) to those perceptual mental images. We see a house or a table that is green. Do we have the green color in the brain? (Fig. 5) Obviously not. The color green is only represented in the mind, or more precisely, this representation is the I/mind.

An organism would not be able to survive in its environment, the macro-EW, without the corresponding self/mind/living being. The mental representations of the objects in the macro-EW are approximately correct, because otherwise the organism would be unable to survive in that environment. Because of this, the noumenal is replaced with the EDWs, and the Kantian phenomena are the objects and processes which exist in the macro-EW in the same way in which microparticles exist in the micro-EW.

Searching for real entities (for the thing-in-itself or noumena) or for the real world is meaningless. The living being/the I corresponds to an organism that could not survive in an environment without this correspondence. The great English philosopher Hume (who awoke the great German philosopher Kant from a dogmatic sleep[16]) would be quite right in denying any causality between the being (the I/self, mind, intentionality, subjectivity, will, etc.) and the organism. However, my theory states that causalities are possible between objects that exist within the same EW, but not between entities belonging to EDWs, because the objects that belong to one EW do not exist for the objects in another EW.

The entirety of mental knowledge is the I which corresponds to the union between the brain and the body situated in an environment. We ought to consider the case of Ramachandran's patient, who had a missing arm. (Ramachandran and Blakeslee 1999, see part 8) The unconscious (implicit)[17] knowledge (that is the self) corresponds to the movement of the arms. The self acquires this knowledge during the development of the corresponding body during the first years of life. After the arm gets amputated, the part of the body is missing, but the knowledge corresponding to that arm is still part of the mind (it is the I/self). The pain (which is the I) reveals the unity of the self, a unity that does not exist within the body. For instance, the brain corresponds to different types of neurons and their relationships, and to the white matter that surrounds the neurons and their synapses, as well as to certain oscillations and chemical reactions, etc. We cannot find any unity inside the brain composed of so many neuronal entities and process. We return to our analogy: a table has this type of unity (it has a surface), which cannot be found anywhere within that huge amalgam of microparticles. (Just like a planet produces gravity, but there are no gravitons.)

The same observation is valid not only for this unity, but also for other features: for instance, the pain is the self/living being. The internal sense or the internal feeling of body or mind is a Ptolemaic epicycle, since there is no external and/or internal side for the I. Instead, there are certain processes of the brain and body which correspond to pain (that is the I/self). The pain does not exist 'inside

16 In the 17th century, David Hume introduced the methodological denial of any kind of causality. Not many years later, Kant declared that Hume's idea woke him up from his dogmatic sleep.

17 The meanings of unconscious, namely implicit, and conscious, namely explicit, knowledge are quite similar, but not identical. For information regarding these and other related distinctions, see Vacariu (2008).

the body' or 'inside the brain'. These notions have been created within the unicorn world and entail either the identity (the identity theory in the philosophy of mind) or the causality between mind and body (Searle 1992).[18]

From my EDWs perspective, both the identity theory (the mind is identical to the brain) and the causality between the mind and the brain are entirely wrong approaches. Many thinkers (including Kant) used the notion of internal feelings. However, all feelings and thoughts do not exist inside the body, being instead the I. Why are they called internal feelings? Internal to what? What is the criteria which makes the distinction between internal and external feelings? Certainly, if a stimulus is external to the body, it produces a feeling which is not internal to the self, but internal to the body. However, pain does not exist inside neither the body, the brain, but it is the I, merely corresponding to the brain and body. The processes take place inside the body, but not inside the I/self/mind/life.

Almost everybody uses phrases like 'inside my mind', but nobody uses 'inside my life'. Why does it seems absurd to say 'inside my life', but it is acceptable to say 'inside my mind'? From my perspective, the mind is identical with life (see the next chapter), so 'inside my mind' has the same status as 'inside my life': they are both completely wrong, since they mix information belonging to EDWs.

Let us suppose you have a headache. Can you claim you have a pain in your brain? Remember color: there is no green color in your brain. Therefore, there is no pain in your brain. The pain is the I/self, not part of the self. What part of the self could we claim as having the pain? The pain (that is the self) corresponds to problems with particular neuronal processes. The pain is neither inside the brain, nor inside the body. Eliminating the internal-external distinction in relation to the I, we realize that all feelings (like perceptions or pains) are the I. All feelings are the so-called unconscious (implicit) knowledge, while our conscious thoughts are the explicit knowledge.[19] Conscious thoughts appear spontaneously, not 'in our mind', but they are the mind, the I.

We saw that the parts-whole relationship would be a very problematic association even for an organism. The existence of a thing depends on the point of view/interactions taken into account. A feeling or a visual representation is not perceived by any internal eye (homunculus); all feelings and mental states are the I/self. If we specify the point of view/interactions, we can determine what objects

18 Today, Descartes' ontological dualism (propounded in the 17th century, it says the mind and the brain are two different ontological substances) is rejected by the great majority of people working in philosophy and in cognitive neuroscience. I emphasize the complete difference between the EDWs perspective and Descartes' ontological dualism. From my perspective, the mind and the brain are not two different ontological substances: the mind is an EW, the brain belongs to the macro-EW. Therefore, the brain does not exist for the mind/self/I.

19 The distinction between implicit-explicit (unconscious-conscious knowledge) is quite used in cognitive psychology. (See Vacariu 2008)

or organisms/cells exist. Again, since any living being (the self or the I) is an EW, it is meaningless to consider that the I interacts with the external environment. Only the brain/the body is situated in an external environment (the macro-EW) and there are interactions between the body and the objects that belong to the macro-EW, but they do not involve the I (or any living being or life). Since the I (or life) is an EW and it is not situated in any spatio-temporal framework, we can say that 'The I/mind/self is' but it would be wrong to say that 'The self/mind exists'.

The idea of internal perceptions[20] is strongly related to the great contemporary debates between the supporters of mental imagery (led by Kosslyn) and those who maintain that mental images have no spatial dimensions (very few proponents, led by Fodor and Pylyshyn). Even before trying to answer this problem, we have to wonder on the way the mind represents visual (not imagery) space. Until now, nobody had doubts: the mind perceives external space. (Fig. 5) This contradicts the rule 'being-perceiving', so it is a false supposition. It is the organism which interacts with the external space, not the self/the I. Again, this is because perceptions are the living being/the I (an EW), while the organism belongs to the macro-EW.

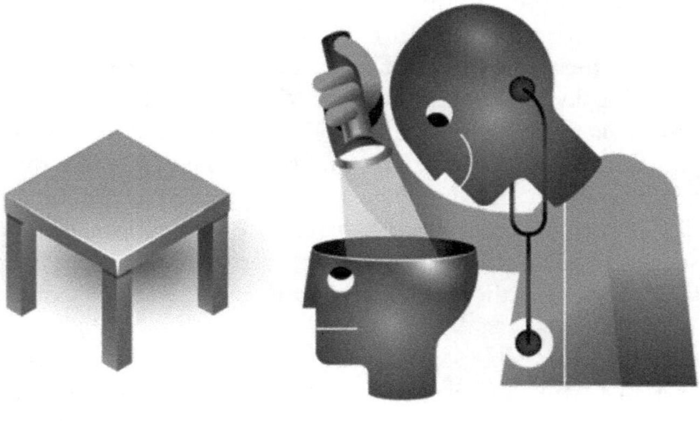

Fig. 5 The mind-brain problem
The patient: „I see a green table standing six feet in front of me."
The doctor: „There is no green color in your brain, there are no feet in your brain, there is no table in your brain." (The mind is neither identical with, nor different from, the brain. The brain and the body belong to the macro-EW, while the mind (the I) is an EW. The brain/body does not exist for the mind and vice-versa.)

20 We have to remember that Descartes imposed notions such as internal and external perceptions in the modern period of philosophy (17th Century).

Because of the appearance of life and because of the evolution of the species, the space has to be somehow represented in the mind, but such mental representations are a kind of virtual space. In the exact same way, we can speak about the representation of the green color in our mind (not in our brain) even if the objects do not have color as such: colors exist only in our minds and correspond to the frequencies of light-waves reflected by objects and received by the eyes, which are parts of the brain/body, the objects and the body all being situated in the macro-EW. We cannot find any space within the mind or life, exactly as we cannot find any green color in our mind or brain. The correlation between space (in which the brain/body is situated) and the mind (in which the space is just represented) is quite wrong.

Many people consider that the mind has spatial features, but nobody inquires about the spatial features of life. In reality, they are the same thing, which is life. From the perspective of the classical notion of objectivity, we have to accept Kant's unproved postulate that the I really is. Extending the notion of the I, we reach the living being or life itself. Nevertheless, since the I has no spatio-temporal determinations, Kant needed to postulate the existence of the I within the unicorn world.[21]

Within the EDWs perspective, we can take this assertion beyond the level of the postulate: the I has no spatio-temporal dimension, and we can prove its existence in other ways. The space of the macro-EW is always determined by its objects. According to Leibniz (a very famous German philosopher contemporary with the famous physicist Newton), space exists as a relationship between objects. Newton believed that space does not depend on the presence of objects, and that it can exist in the absence of any object, therefore describing an absolute space and an absolute time. However, Einstein showed that the absolute space and the absolute time are meaningless, everything (including space and time) being relative to something and not existing independently of everything (absolute existence). (See Chapter 8 of this book)

We do not have a representation of empty space, void of all objects. We can have mental representations of objects and certain relationships between them (and these representations are the I), but such representations simply correspond to certain biological mechanisms of an organism that interacts with a specific external environment. (Fig. 6) Even if the macro-EW has spatial dimensions, space has to be suspended in the mind, where it is merely represented. If the mind (self) had a spatial aspect, it would be decomposable. This decomposition is not possible (can you decompose yourself?), so the mind has no spatial dimensions.

21 However, during Kant's times, there was no knowledge of the quantum world. Interestingly, Kant did not deal with the mind-body problem initiated by Descartes more than one century before.

Again, I return to the person watching a TV screen: the eyes receive certain electromagnetic waves that are transformed into electrical and chemical signals in the sub-cortical and cortical areas. There is a representation of that TV in the mind, but there is no TV in the brain. The neurons, the white matter, the whole body are situated in space, but there is no space inside the mind. The TV screen is 2D (it has two spatial dimensions), but through correspondences the I perceives the scene from the TV set as being in 3D. The mind has a kind of 'representation of space' (virtual space) necessary for the corresponding organism to survive in its environment. Many people argue that space exists in the mind because the brain has a spatial extension. This is an illicit extension that breaks the Kant-Carnap rule.

Fig. 6 The self (the I, the mind)
A body (brain) is on a beach. There are some interactions between the sand, the ocean, the light, the body, (arms, legs, eyes, brain), the chair, etc. This person has an image (representation) of her body and of the beach (ocean) where it is placed. This mental image is the I (self) and it corresponds to the body (brain) and to its interactions with the environment. The self does not interact with the external environment, because it does not exist for the external environment (the macro-EW, in this case, the beach and the ocean); the external environment does not exist for the I. This image (that is the self) does not exist for any neuron of the brain, it is not in the brain, it does not exist for the eyes. The brain/body does not exist for the mind/I.

Space is only represented in the mind, precisely the way color is, and both of them are nothing more, nor less than the I. There is no direct relationship between the external space or color and the representations of space or color. I repeat, the representation of space is something completely different from real space or the area occupied by the brain. There is no localization/representation of mental space in our brain. Researches working in the domain of cognitive neuroscience are trying to localize the mental representation of an object in the brain, but only the object itself is in a spatial (-temporal) framework that belongs to the macro-EW. Moreover, the brain is in a spatio-temporal framework in relationship with other entities, but for the neurons that compose the brain, the brain does not even exist.

The error concerning the 'perception of an external world' is one the main reasons why human beings created the unicorn world. People working in cognitive science have completely ignored Kant's main idea that the external world (the representations of the external world) is the self (Waxman 1995). 'The self is' does not mean that the self/the I interacts with something else (only entities that exist interact). 'The self is' means that the I is an indeterminate individuality, i.e., it does not have spatial determinations. (Fig. 6) Determinations demand interactions (that require a spatio-temporal framework), indeterminations exclude interactions, but require an indeterminate identity (an identity without spatial dimensions, exactly like the I/self/mind). To exist means to interact or to be determinate, as any object or organism exists. To be means not to interact, or in other words, to lack spatial determinations, to be the self/mind/life or an EW. We have to change the notion of 'objective reality' completely. All entities (objects, organisms, minds/selves) and all EDWs have the same objective reality. Space and time are not the criteria for objective reality. Hume was right to be skeptical about causalities between many entities. This skepticism, clarified by the EDWs, changes the status of existence/being.

3
Epistemologically Different Worlds (EDWs)

The hyperverse is an abstract notion, ontologically-epistemologically speaking, and it represents the sum of all EDWs. Ontologically, independently of our existence/observation, the entities of a particular EW do not exist for those in another EW. Epistemologically, using no more than one set of conditions of observation, human beings are unable to observe (indirectly, through correspondence) entities from more than one EW at any one moment; this is due to the nature of attention/consciousness, which is a serial process.

We have to keep in mind that all perceptions/observations are indirect for at least one reason: any mental perception is the I, which is an EW in itself – and no EW exists for any other EW. For instance, any mental representation or mental image of a macro-table which belongs to the macro-EW is the mind-EW. This mental representation corresponds to the physical processes taking place in the eyes and the brain, which similarly belong to the macro-EW. In order for the hyperverse to have an ontological status there needs to be an entity which can interact with (observe) several entities from EDWs at the same time. We will call this a hyperentity. Does such a thing exist?

Human beings are not hyperentities: a person cannot observe (through correspondence, not through direct access) entities that belong to two EDWs at the same time. Even if it were possible for our consciousness/attention to run parallel processes, we are unable to be in two conditions of observation, using two types of mechanisms of perception, at the same time. We cannot do multiple observations which involve multiple apparatuses/mechanisms of observation/perception. In reality, a hyperentity would need two conscious minds to observe two EDWs. Since the mind is the living being/the I that is an EW, a hyperentity would need to be two EDWs at the same time, which is clearly a hyperontological contradiction.[1]

In order to observe (indirectly, through correspondence) the entities in a particular EW, we need certain conditions of observation. Observation is a unidirectional process, since some of the entities we observe indirectly, through

1 Until this point in the book I have used the phrase 'ontological contradiction', but a more precise way of naming it is 'hyperontological contradiction', because it involves entities belonging to at least two EDWs.

correspondences, are not in the same macro-EW as our brain/body, which correspond to the I. For instance, with the help of an electronic microscope (a macro-object), a researcher observes (indirectly, through correspondence) an electron that belongs to the micro-EW. The electron does not observe or interact with neither the brain/body, nor with the mind of that researcher. For the electron neither the human brain/body, nor the human mind exist; the electron interacts with an amalgam of microparticles which corresponds to the human body. Moreover, an electron does not observe/interact with the electronic microscope, which is a macro-object, but only with an amalgam of microparticles that correspond to the electronic microscope.

The processes we use to observe entities belonging to EDWs are always indirect and happen through correspondence, even in the case of macro-objects, not just for micro-objects. The rule has to be very clear: in order for two entities to interact (to observe each other), they have to belong to the same EW. Each entity can observe/interact with another entity only if they belong to the same EW. With the exception of the human organism, no entity can change its conditions of interaction/observation so that it may observe something indirectly, through correspondence. All entities except human beings have their ontological/epistemological limits imposed by their conditions of interaction.

Even if we change the conditions of observation, or if we use the same conditions with modified parameters, we will still (indirectly) observe different entities belonging to the same EW. To change our conditions of observation we need to cross an organizational threshold so that we may shift our powers of observation from entities that belong to one EW to entities that belong to the same EW. For example, let us discuss the whole-parts relationship. As I emphasized in Chapter 1, the whole does not exist for the parts, nor do the parts exist for the whole. Moreover, a person cannot observe (pay attention to, be conscious of) both the whole and the constituents at the same time. Therefore we have to reject the existence of the whole and the parts in the same place at the same time. However, if we change the conditions of observation and if we pass an epistemological-ontological threshold, we can indirectly observe entities that belong to EDWs, shifting from entities and processes belonging to one EW to those belonging to another EW.

We have to keep in mind the fact that in order to observe entities which belong to a new EW, we have to construct the conditions of observation which will allow us to interact indirectly with the entities in that EW. Following the ideas of the great physicist Bohr (see Vacariu 2005, 2008), any new apparatus of observation is a macro-tool constructed and manipulated by our hands/body (including the brain) which correspond to the mind. In Kantian terms, within the EDWs perspective, the conditions of possibility[2] for our tools of observation should reflect at least in part (in certain determinations) the conditions of possibility for the interaction of a set of organisms (and/or objects) that belong to an EW (see Vacariu 2008). We can become aware of EDWs only through the 'hyperontologization of epistemology'. The ontologies of EDWs have become epistemologies and vice-versa. Nonetheless, these ontologies are not different ontologies but epistemologically different ontologies which represent the hyperontology of the hyperverse (this is an abstract expression, since the existence of a hyperentity with two minds is impossible).

The EDWs perspective is really a new framework of thinking that requires a new language. As an abstract notion, the hyperverse creates the semantic framework necessary for the understanding that EDWs are but not exist (only objects exist). Each set of interactions constitutes the surface[3] of the entities that belong to an EW. Thus, the surface somehow mirrors the EW in which it determines the interactions which constitute the object. The I only represents the external world, which is only external to the brain/body, not to the I, since any mental representation is the self.

Ontologically speaking, there are no 'entities inside an object', because that would break the whole-parts rule: those parts correspond to the object. It is an ontological contradiction to consider that the parts of a car (as independent entities), and the car (as a whole) exist in the same place at the same time. Again, there are no two sets of objects that ontologically exist in the same place, at the same time. Either the car, or the set of its parts can exist in a certain place at a certain time. Therefore, the phrase 'inside an object' is merely an abstract notion. As we saw above, it is also meaningless to talk about an internal-external relationship when it comes to the I/mind/life. Life is neither external, nor internal to a body, it does not emerge or reside in it; instead, it corresponds to the body. This is also the case with a car or a table, which we cannot claim exist in the same place, at the same time with their parts, or with the microparticles and the relationships between them that correspond to those objects.

As a whole, an EW has no (spatial) boundaries/limits, otherwise it would interact with another EW, which is impossible. In fact, space exists only among objects, but not among EDWs which do not exist for one another anyway. What entity could be found outside the macro-EW? This is a meaningless question, because in this case outside would come to mean 'outside space' - and 'outside an EW', space does not exist. This is another reason why we cannot assume that the

2 The conditions of possibility refer to the framework that is necessary for the existence of an entity.

3 As far as natural objects (such as stones) are concerned, the distinction between the surface and the interior is meaningless. For the self or the I, the distinction between the exterior and the interior is similarly wrong.

mind-EW (the life-EW) has spatial dimensions (in cognitive neuroscience, this is called 'spatial cognition'). Therefore, any EW is indeterminate exactly as the I/self/life is an indeterminate individuality. If an EW were determinate, an object could interact with that EW. Then that object would be part of that EW and not outside of it, so the object does not interact with the EW, but it is part of it.[4] Those interactions would need a spatio-temporal framework, but there is no outside (or inside) for any EW. As I argued above, space exists only among objects belonging to the same EW, so it is meaningless to ask if our world is spatially infinite or not. We can only say that the entities/objects of an EW are within a spatio-temporal framework (that is, their relationships presuppose a spatio-temporal framework). We saw above that any EW appears from and disappears into nothing[5] (which has neither determinations, nor dimensions).

What does the expression 'epistemologically different' actually mean? Obviously, it does not mean the same thing as 'ontologically different', which refers to ontologically different substances or different types of matter. There is no ontological meaning for my expression. The difference is neither ontological (as Descartes believed was the case for the mind and the body/brain), nor linguistic (the way Carnap, a famous philosopher belonging to Vienna circle, believed it to be). The notion of 'epistemological difference' imposes certain hyperontological limits related to the limits of each entity in any EW.

'To exist' or 'to be' means to have certain limits, not necessarily spatial ones, which entail determinations. Even the living being/the I as an indeterminate individuality has limits (the self is not infinite in any way) or, more precisely, it has certain epistemological-ontological limits. The notion of 'epistemological difference' assigns to each class of entities the same epistemological abilities that man has, i.e. observation/interaction (within the EDWs perspective, these are epistemological-ontological abilities). As we saw in the previous chapter, if I were a planet (or an electron), I would interact with another macroparticle (or with microparticles). If I were a cell, I would interact with the environment specific to a cell. However, the living being (life) that corresponds to a cell does not interact with anything else, since it is an EW.

The expression 'epistemologically different' eliminates many of the speculations (Ptolemaic epicycles) that philosophers and scientists have developed over centuries. It eliminates the ontological-epistemological contradictions typically

4 This observation is valid even for Everett's 'many worlds' or 'parallel universes'. If there are two 'parallel universes' (a notion used by actual physicists, which is in fact contradictory), both universes can interact, at least theoretically, therefore they are in the same unique macro-EW. Therefore, the two universes would be not in fact two universes, but only two parts of the same macro-EW.

5 The correct notion is not nothing but hypernothing. (For more details about this notion, see Vacariu 2011)

available within the unicorn world. The human organism needs to change its conditions of observation in order for a human being to observe (indirectly, through correspondences) certain epistemologically different entities that belong to EDWs.

Now we can clearly understand the expression 'epistemological-ontological'. Changing certain conditions of observation (the difference between them being an epistemological-ontological threshold), the I observes EDWs (indirectly, through correspondence). In other words, the threshold is an epistemological-ontological one between entities that belong to EDWs. We think that the knowledge we have about the world is certain, but many parts of it are false. These distortions, instead of representing the truth about certain entities (objects or minds/lives that really exist or are), they present the pseudo-relationships (causalities or not) between them. In the position of the dictator-observer, man has imposed the tyranny of the unicorn world.

From the point of view of the human being, it seems that all entities belong to the same spatio-temporal framework. From the point of view of another entity, an object can observe only the other objects that interact with it. In general, the interactions between certain objects take place in a spatio-temporal framework. The framework of an object (for instance, that of a microparticle) is not the same as the spatio-temporal framework of a biological human organism, since the microparticle does not observe/interact with the macroparticle. Therefore, microparticles and macroparticles are in epistemologically different spatio-temporal frameworks. (Fig. 3) This is the main reason why we have to reject the idea that all entities are within the same spatio-temporal framework, i.e. the unicorn world. Obviously, assuming that everything exists in same spatio-temporal framework can be (pragmatically) helpful in our daily life. However, in science (and in philosophy) the fundamental problems require the EDWs paradigm.

From the human point of view, it would seem that the number of EDWs is not too large. If we extend the conditions of observation/interaction to all entities, however, the number of EDWs increases considerably. If we accept that being is and that it corresponds to an organism, we have to reject the notions of levels, attributes, supervenience, composition and elementary particles. Being corresponds to an organism, therefore we have to hyperontologize all classes of entities that do not interact or emerge or are identical (those that have an epistemological difference). An entity needs to have a unity that represents its identity, even in the case of an indeterminate individuality such as the self/mind/life. In this context, I will introduce the next proposition, the principle of hyperontologization:

(11) The I is, therefore EDWs are.

The unity of the I/self/life and the unity of a planet have an ontological character: both are/exist. If we were just decomposable organisms, or if the I lacked unity, we would be unable to acknowledge the existence (being) of EDWs and the I would not be an entity. Only the I (the self with its unity) is able to discover the being (existence) of EDWs. The relationships between the mind and the brain (between life and an organism, or between the whole and its parts), such as identity or emergence, are rejected.

Reductionism completely fails to explain the relationships between the notions and the theories of various branches of science; emergence is a pseudo-notion, so the mind does not emerge from the brain. (See Chapter 4 of this book or Vacariu 2008) However, the mind/living being is, while an organism or cell exists. Within the unicorn world, they cannot both be and exist, since that would be an ontological contradiction. Therefore, the only alternative is to have a correspondence between the living being (life, mind) and an organism/cell/brain. However, without the existence of an organism in the macro-EW, the living being would not be another EW. Within the unicorn world, we prefer to use either notions like emergence or causality to describe the relationship between different pairs, or fundamental particles. If fundamental particles existed (and a large number of physicists believe that they do), the self/I/life would not be at all.

Based on implicit/unconscious knowledge, any conscious/explicit thought appears spontaneously in the mind. Let us suppose that you are in a conversation with a friend and that you talk uninterruptedly for two minutes. You are not reading anything, you are simply telling the other person what you think about the corruption in your country. Being quite involved in that conversation, you speak rapidly and even make a few grammatical mistakes in your native language. The questions I have for you are: do you mentally perceive each sentence that you pronounce before you actually pronounce it? Do you have internal eyes reading the sentences that you pronounce? Are you aware/conscious of each sentence before you pronounce it? Clearly, the answer to all these questions is negative.

Generally, any sentence appears spontaneously in your mind (it is the I), whether it is pronounced or not. The notion of spontaneity (which is essential for Kant[6]) creates the space necessary for the living being (the I) and representations (that are the I) to appear. It reflects the unity of representation and the unity of the mind. It corresponds to the most active neural patterns/processes that pass a

6 „The synthesis of apprehension that involves the imagination needs the same spontaneity as the synthesis of apperception that implies the understanding. And this spontaneity brings combination into the manifold of intuition. But this combination of the manifold of intuition determines the unity of representations. Thus, spontaneity would determine the unity of representation that, as I said above, is given by formal intuition.“ (Kant's „Critique of Pure Reason“) For more details about Kant's transcendental philosophy, see Vacariu (2008).

certain threshold of activation. The unity is Kant's essential notion of synthesis necessary for thinking (which includes perception). If, for Kant, synthesis is a transcendental process, for me synthesis is the implicit/unconscious knowledge, it is the conditions of possibility for explicit/conscious knowledge, it is the I.[7] If Kant writes „Even for space as an object, we need the unity of the combination of the manifold of a given intuition", for me space does not exist in the mind/the I - the representation of space might, but not space itself. 'Inside the mind' (which as I have mentioned is a wrong notion), color is in the same situation: it does not exist as such, but it is merely represented. Spontaneity is the determination of the I.[8]

As a whole, the I is an indeterminate individuality (implicit knowledge), but the spontaneous appearance of an explicit/conscious representation determines the I to be in a certain state. Thus, we have to make another distinction for the I, between the explicit and the implicit states. Spontaneity is indeed 'a determination of my existence' (see the footnote below) but this determination has to be explicit[9], since 'the I is' and might or might not have determinations. More precisely, these determinations are the I/living being. Spontaneity explains explicit/conscious knowledge. Thoughts could not appear out of nothing, they are parts of the implicit/unconscious knowledge (memory, etc.) which become explicit/conscious thoughts. Only any EW (including the living being) appears out of nothing. Conscious/explicit thoughts appear spontaneously from the unconscious/implicit knowledge (that is the I). The indeterminate individuality (of the living being) would be the condition of possibility for this type of spontaneity. The Kantian transcendentalization that is ontologically loaded in 'immanent thinking' becomes 'immanent interactions'.

Even if in my 2008 book I provided ample proof that the EDWs perspective is, in a way, an extension of Kantian philosophy, I will continue to develop this analysis here. For Kant, the representations of the external world are the self. The body/brain exists as an entity in the macro-EW, while the mental representations of the body are the I. Kant wanted to construct the philosophical fundaments of Newton's theory in order to explain the world. Today, under Einstein's influence, Friedman felt the need to relativize Kant's theory. (Friedman 2001, see Vacariu 2008)

7 As explained in Chapter 2, the I is both implicit/unconscious and explicit/conscious knowledge.

8 Commenting B158, for Pippin, spontaneity seems to be the „determination of my existence". (Pippin 1997, p. 34) Spontaneity comes from the self, which is the being with an indeterminate individuality. However, I quoted the last sentence from B158: „But it is owing to this spontaneity that I entitle myself as intelligence." (See also A546–7/B574-5, in Pippin, p. 34, in Vacariu 2008) To put it in the language of EDWs, we are talking about explicit, not implicit, knowledge.

9 When I claim that the living being (the I) is an indeterminate individuality, I refer to it in general, regardless of any spontaneity.

One fundamental element in Einstein's special relativity theory is the postulate regarding the constancy of the speed of light in relationship with any point of reference that, according to Friedman, acquires the status of „coordinating or a priori constitutive principle".[10] To extrapolate Kant's idea, principles of this sort define the „the fundamental spatio-temporal framework of empirical natural science". (Friedman 2001, p. 43) Each scientific theory has certain a priori constitutive principles that define its proper space of empirical possibilities (Friedman 2001, p. 84) or conceptual frameworks that „define the fundamental spatio-temporal framework within which alone the rigorous formulation and empirical testing of the first or base level principles is then possible". (Friedman 2001, pp. 45-6) (for more details, see Vacariu 2008)

For Einstein, the coordinating principles constitute a new framework for space, time, and motion (Friedman 2001, p. 107) and therefore all empirical laws have constitutive meaning only in the framework created by a priori constitutive principles. Even the individuation of entities requires such conceptual frameworks. This is necessary not only because the entities in motion belong to a certain spatio-temporal framework, but also because „the knowledge of physical rigidity presupposes the knowledge of forces acting on the material constitutions of bodies". (Friedman 2001, p. 110) From the EDWs perspective, the interactions individualize (constitute) the entities within a spatio-temporal framework and the rigidity of physical objects is just their surface.

What does 'practically rigid bodies' mean for Kant? In order to describe forces, Einstein used geometry. Essential for the EDWs perspective is Friedman's footnote on page 55 about Einstein, who adopted a perspective on the relationship between a necessary geometry and entities as 'practically rigid bodies' which ignores microphysical forces. (Friedman 2001, p. 114) We simply need strong reasons to ignore the essential forces within the world. The only solution to ignore such forces is the introduction of EDWs. Obviously, analyzing phenomena that belong to the macro-EW, we can ignore the microphysical forces (that belong to the micro-EW), since micro-objects and their forces (electromagnetism[11], weak and strong forces) do not exist for macro-objects and their force (gravity) and vice-versa. Without EDWs, we appeal to a postulate (which by definition is not proved) that brings us to the realm of so-called 'empty notions'.

10 Einstein transformed the light principle, which was an empirical principle for Newton, into a constitutively a priori one. „Einstein elevated an empirical law to the status of a convention or to the status of a coordinating or constitutive principle". (Friedman 2001, p. 88, in Vacariu 2008, p. 303; for more about Einstein's relativity and EDWs, see Chapter 8 of this book)

11 I mention them here in a simplified manner, even if in the last chapter we have seen a detailed analysis of the fact that even waves and microparticles belong to EDWs.

Within the context of the EDWs perspective, it is important to answer the following question: 'What was there before the Big Bang?' Most physicists would tell us that this question is meaningless for the only reason that they do not have any plausible (scientific) answer. From my point of view, this question has a plausible (philosophical) answer. I believe that there was an EDW (or maybe that EW still exists), which I will call the 'pre-Big Bang EW'. (See also Vacariu 2012) However, the micro-EW (or the macro-EW) did not appear from the 'pre-Big Bang EW'. There are no causalities between any two EDWs. The idea of any kind of causality between ED entities that belong to EDWs is meaningless. Obviously, there are some correspondences, but we cannot speak of causalities. Any EW appears from and disappears into nothing. Then what is the role of that 'pre-Big Bang EW'?

There are some correspondences between ED entities and processes that belong to the pre-Big Bang EW and the micro-EW (this is allegedly the EW that first appeared after the Big Bang). Again, 'what was there before the pre-Big Bang EW?' Was there another EW? Then there could be an infinite chain of EDWs. How can we stop the expansion of this infinite chain? Moreover, how could we avoid having a theoretically small or big infinity? In the case of the small infinity, imagine dividing a table in infinite parts, while in the case of the big infinity we can imagine traveling in infinite space and time.[12] Within the unicorn world, nothing could stop us from thinking of such infinities. We can only rule them out by using the EDWs framework.

Before the pre-Big Bang EW there might have been another EW, and before this EW there could have been another EW and so on, but we do not have an infinite chain of EDWs. We can stop this infinite chain of EDWs by assuming that, in this chain of EDWs, there was an EW (let me call it EW0) that had no spatio-temporal framework. It is possible for this EW0 to be because there is another EW that lacks spatial dimensions (the mind-EW) and some entities exist without a temporal coordinate (e.g., photons) that belong to a particular EW (the micro-EW). Therefore, if we have an EW without space and time, the question 'What was there before it in the chain of EDWs?' is rendered meaningless. If we talk about the EW0, the questions referring to the spatio-temporal framework of the entities belonging to this EW are meaningless.

12 These infinities remind us of Kant's antinomies of space and time, atomism, causality (freedom). From my point of view, there needs to be an alternative to all of Kant's antinomies (including the antinomy concerning God). With EDWs, we avoid the notion of the world and therefore we rule out any kind of infinity (including the existence of God). (Using the EDWs perspective, I demonstrated that the notion of God – like those of world and infinity - is meaningless. On my webpage, see my presentation „God died a long time ago. How can we rule out the infinite?" from the Symposium „Theism versus Atheism", September 2012, Bucharest University)

Therefore, we cannot divide a table in infinite parts, nor can we theoretically travel in infinite space and time because at 'a certain moment'[13] we move into the EW0 that has no spatio-temporal framework. What is important to remember from this example is that, within the EDWs perspective, we rule out any kind of infinity. Notions such as the world, infinity and many others have been created by the human mind within the unicorn world. It is now time to renounce invented notions like these, which have always created great problems for the human understanding.

Now we can create a table with all the dichotomies analyzed in Chapter 1 (non-living objects) and Chapter 2 (entities such as organisms and cells which correspond to the mind/the I/life).

Entities in the micro-EW	Entities in the macro-EW	Entities in the macro-EW	Entity = EW
Microparticles	*Macroparticles*	*Macromacroparticles*	-------
Microparticles	*Macroparticles*	*Cells*	*Life*
Microparticles	*Macroparticles*	*Organisms*	*Life*
Microparticles	*Macroparticles*	*Brain/body*	*Mind (I)*

The elements in the first column (the microparticles) belong to the micro-EW. All entities from the second and third columns (macroparticles, cells, organisms, brains/bodies) belong to the macro-EW. The entity from the last column is an EW. The relationships between the elements of each row is that of correspondence. For instance, in the first row, the microparticles (micro-EW) correspond to the macroparticles (macro-EW) that correspond to macro-macroparticles (macro-EW). In the second row, some microparticles (micro-EW) correspond to (they do not compose) some macroparticles (macro-EW) that correspond to cells (macro-EW). In the third row, certain macroparticles from the second column correspond to an organism on the third column, which in its turn corresponds to life. In the last row, some microparticles correspond to some macroparticles that correspond to a body that corresponds to the mind/the I.

On the first row of the third column, we have 'macro-macroparticles'. We will remember the example I used in Chapter 1: an automobile (or a table) is a macro-macroparticle, by which I mean to say that it corresponds to some macroparticles. Again, according to the 'whole-parts' rule, the car (the macro-macroparticle) does not exist for its macro-particle components, such as the engine, doors, wheels,

13 This notion is quite improper.

windows, and other parts (and vice-versa), even if both macro-macroparticles and macroparticles exist in the same macro-EW, but not for the same interactions and not at the same time. Moreover, the car (or the table) is not composed of microparticles (that belong to the micro-EW), but corresponds to these microparticles. The same rule is valid for the relationship between an organism and its cells, or between a cell and its components (the membrane, the DNA and RNA, molecules, enzymes, etc.). All human bodies (with their brains) and the bodies (with brains or not) of all animals, as well as plants, cells, and all macroparticles belong to the same macro-EW. However, cells do not compose an organism, they correspond to it. As we saw in previous chapters, the brain/body does not exist for its neurons/cells, while the neurons/cells do not exist for the body, so it is meaningless to claim that the neurons/cells compose the brain/body. If the cells of a body (or the macroparticles, or microparticles that correspond to the body) scatter instantly in a room, its cells (or the macroparticles, or the microparticles) still exist, but the body does not exist.[14] Essentially, do not try to find the relationship between the whole and the parts: since the whole does not exist for its parts, their relationship does not exist.

There are microparticles and macroparticles that do not correspond to any cell or organism. However, any cell/organism that exists in the macro-EW corresponds to a particular set of macroparticles and microparticles (that belong to EDWs). Since the entities I have inscribed in columns are or belong to EDWs, the relationships between these entities are merely those of correspondence. A mind is an entity/process that is an EW and corresponds to a body (brain) that corresponds to an amalgam of cells that correspond to a set of macroparticles (all the latter elements belong to the macro-EW) that corresponds to a set of microparticles (that belongs to the micro-EW).

14 Obviously, all these cases revolve around the paradox of the „ship of Theseus". I believe that Wikipedia explains the matter rather well: „The ship of Theseus', also known as Theseus' paradox, is a thought experiment that raises the question of whether an object which has had all its components replaced remains fundamentally the same object. The paradox is most notably recorded by Plutarch in Life of Theseus from the late 1st century. Plutarch asked whether a ship which was restored by replacing each and every one of its wooden parts remained the same ship. The paradox had been discussed by more ancient philosophers such as Heraclitus, Socrates, and Plato prior to Plutarch's writings; and more recently by Thomas Hobbes and John Locke. There are several variants, notably 'grandfather's axe'. This thought experiment is 'a model for the philosophers'; some say, 'it remained the same', some saying, 'it did not remain the same'." (http://en.wikipedia.org/wiki/Ship_of_Theseus) From my point of view, the ship of Theseus is a wonderful Ptolemaic epicycle! It is no surprise that nobody could solve it after more than two millenniums: the question is meaningless since the ship does not exist for its parts and vice-versa. Again, I repeat, with the EDWs perspective, many paradoxes and problems in philosophy and science become merely 'linguistic games' created by the human mind over several millenniums. Such linguistic games have nothing to do with reality, i.e., with EDWs. During the last century, philosophy has become full of such sophisticated linguistic games elaborated by philosophers who specialized on a particular topic or area of philosophy.

Finally, for a better understanding of the EDWs perspective, we can make an analogy between the following two pairs: microparticles-planet and brain-mind. The microparticles that correspond to the planet do not produce or cause its existence, while the planet does not emerge from the microparticles. Moreover, the microparticles are not identical with the planet since there are different forces for the microparticles (quantum forces) than there are for the planet (gravity). The features of the microparticles and those of the planet differ. The same relationship is valid for the second dichotomy of the brain and the mind: the brain does not produce the mind (Searle 1992); the mind does not emerge from the brain.[15] The mind is not identical with the brain since the mind and the brain have completely different features. The mind is an EW, the brain belongs to the macro-EW exactly as microparticles and macroparticles belong to EDWs.

15 There are different kinds of emergence (see Chapter 4 of this book; for more details see Vacariu 2008). Many people (both philosophers and scientists) believe in one type or another of emergence but all kinds of emergence are pseudo-notions.

Part II

Applications: Philosophy, Cognitive (Neuro)science, Biology, and Physics

4
EDWs and the Philosophy of Mind

In this chapter, I will use the EDWs perspective to analyze different notions from the philosophy of mind produced within the most powerful and dominant old paradigm in the history of human thought, the 'unicorn-world'. Due to the paradigm of the unicorn-world, the major problems in the philosophy of mind are in fact pseudo-problems. In this chapter I will analyze some notions (levels vs. reduction, emergence) which involve the main directions in the philosophy of mind: dualism, identity theory, eliminativist materialism, and non-reductive physicalism.

4.1. Descartes' dualism

In the 17 century, Descartes' dualism constituted the foundation of one essential direction of Western thinking, namely, rationalism. He officially introduced the mind-brain/body problem in philosophy. In a very religious period and working in a thought framework specific to that period, Descartes believed that the mind and the brain were two ontologically different substances. This is his dualism: two ontologically different substances that form a person.

It has always been clear that the mind can act on the body (for instance, the mind can cause the movement of an arm). Therefore, the main problem for Descartes (for dualism in general) was the union/causality between the mind and the brain/body, that is, the union between two ontologically different substances. If two substances are ontologically different, what can unify them or how can they be unified? As it is very well known, even Descartes' contemporaries (Regius, Gassendi, Arnauld, Princess Elisabeth, etc.) criticized his dualist perspective. I think that the source of Descartes' error is that he created his dualist approach in the preexistent paradigm of the unicorn-world perspective. I will offer a short presentation of Descartes' philosophy, then I will criticize it within the EDWs perspective.

The most important notions of Descartes' philosophy are (1) the subject (the I as a 'thinking thing'), (2) two kinds of perception, (3) two kinds of substances (on the one hand: God, angels, and the mind, while on the other: the body, animals, and inanimate things), (4) one world and (5) the relationships between perception and substance, between the mind and the body, between the I and the world.

Descartes' first step was to introduce the demon hypothesis[1] that our knowledge of the existence of all external things and of the body is under doubt because it is produced by our senses or by our dreams. (Descartes 1994, Meditation I, 74–5) All external objects and (parts of our) bodies belong to „corporeal nature in general and its extension". (p. 76) This kind of knowledge referring to 'composite objects' (like those of physics, astronomy, medicine, etc.) has a doubtful character. On the contrary, sciences which deal with simple and general objects – like arithmetic, geometry, etc. – are built on indubitable and certain knowledge. (p. 76)

Thus, Descartes' second step was to find something that we know exists without any doubt, representing the certain and indubitable Archimedean point: the I. If the I can be deceived, it follows that the I exists. The conclusion is: this proposition (pronunciatum), that I am, that I exist, is necessarily true each time it is expressed by me, or conceived in my mind. (p. 80) As a thinking thing, I has different functions (or properties), such as doubting, understanding, denying, willing, sensing and imagination. (Descartes 1994, p. 82) „Je pense, donc je suis" (I think, therefore I am) is „the first principle of the philosophy I was seeking" (Descartes in AT VI, 32; CSM I, 127 in Fowler 1999, p. 63)

Here is one of the most quoted passages from Descartes in which Wilson identifies the epistemological argument:

> Because I know that all that I clearly and distinctly understand can be brought about by God as I understand it, it is enough that I can clearly and distinctly understand one thing apart from another, for me to be certain that one is different from another, because they can be placed apart at least by God; and it doesn't matter by which power this is done in order for us to judge them to be different; and thus, from this fact that I know I exist, and that meanwhile I notice nothing else to pertain to my nature or essence, except this alone that I am a thinking thing, I rightly conclude that my essence consists in this one [thing] that I am a thinking thing. And although probably (or rather as I will afterward say, certainly) I have a body, which is very closely conjoined to me, because nevertheless on the one hand I have a clear and distinct idea of myself, in so far as I am only a thinking thing, not extended, and on the other hand I have a distinct idea of body, in so far as it is only an extended thing, not thinking, it is certain that I am really distinct from my body and can exist apart from it. (AT VII 78 [CSM II 54] in Wilson 1998, p. 189)

What do we understand from this excerpt? We can distinguish the following string of ideas: I can perceive clearly and distinctly two different substances, the mind and the body, the conjunction between mind and body (see below) and the relation between essence and existence. I will restate the Cartesian argument as to why the mind is distinct and different from the body: 'because they can be placed apart at least by God; and it doesn't matter by which power this is done in order for us to judge them to be different'. We judge those substances as being different

1 Descartes imagined a godlike being of absolute evil, a demon, deceiving him and twisting reality in such a way that it was impossible to know whether anything he perceived was truly real.

because God has created this power in us. Even if the I, as an Archimedean point, is the point of the reconstruction of the world, for Descartes, God remains the foundation of his edifice. In fact, God and faith represent for Descartes (of course, before Kant) the leap of human beings beyond their perceptual/conceptual limits.

Descartes emphasizes the fact that the mind is indivisible and the body is divisible. (Descartes, Sixth Meditation) Moreover, in the Synopsis, he insists on the difference between the mind and the body. (Fowler 1999, Ch. 8) This conclusion is confirmed in the same Meditation (the sixth) by the fact that we cannot understand a body except as being divisible, while by contrast we cannot understand a mind except as being indivisible. He claims that we cannot conceive of half of a mind, while we are always able to conceive half of a body, no matter how small it is. This leads to the realization that the nature of the mind and that of the body are more than simply different, being in some way opposite. (AT VII; CSM II 9–10, in Fowler 1999, pp. 274–75)

The opposition between the nature of the mind and that of the body is reflected by their opposite properties, materiality and immateriality. (Fowler 1999, p. 276) Thus we are aware of and we understand these properties (which belong to the ontological level) only through our capacity of perceiving the two substances as clear, distinct and complete things. If, by using these processes of perception, we identify two different sets of properties which belong to different entities, then evidently the substances that correspond to those objects exist.

One of the fundamental issues I have mentioned needs to be kept in mind: Descartes' approach is grounded in the pre-existing framework (paradigm) which has dominated human thought throughout history, the unicorn-world perspective. I believe that within this framework there is a key element which reveals one of the most important errors: the postulation of one world, which represents the single ontological world where everything has been placed. In fact, there are two factors (psychological and religious) that have shaped the unicorn-world. Psychologically, we have the sensation, the feeling that both the mind and the body exist without any doubts. Each of us is the I that can perceive his/her body, so both the I and the body have to exist. The religious argument is related to the existence of God. If we believe in his existence, then, even if he has no spatiotemporal dimensions, he has to be present in the whole universe. God's power implies the existence of only one Cosmos or Universe. God exists everywhere, and this everywhere implies the existence of a single Universe.

In Descartes' framework we can notice another bidirectionality between two notions: the I and the world. As a thinking thing, the I perceives/conceives the self as a single entity (being). But at the same time, the I perceives/conceives the external world as one world. The nature of our thinking is reflected by this bidirectionality: one being presupposes one world, or an unique external space. Descartes' primitives are one I, two perceptions/conceptions, two kinds of substances, and one world. One person in one world is composed of two substances, mind and body. Thus, the question is: how is it possible to situate two completely different substances within the same entity?

Descartes' main problem is the unity of mind and body. Even if the mind and the body are separate substances with opposing characteristics, their unification is necessary because of the uniqueness of both the I and the world. Descartes was aware of the difficulty of the mind-body union and was actually unable to solve it. Several of the thinkers of that age, among which Regius, Arnaud, Gassendi and Princess Elisabeth, greatly criticized his idea of the union of mind and body, sensing that the problem has no logical answer. Descartes tried to justify the union between mind and body almost fanatically.

I think that there are two lines of arguments that support Descartes' belief: religious ones and philosophical ones. As Fowler emphasizes in the conclusion of his book, Descartes, preserving a traditional relation between doctrine and philosophy, rejects Regius' alternative of the 'double-truth option', i.e., of separating the truth of revelation from the truth of reason. Reaching the stage in which he was aware that the unity between mind and body couldn't be proved scientifically or philosophically, Descartes pronounced, „the union of mind and body is a reality which escapes philosophical discourse". (Descartes to Elisabeth, 21 May 1643 in Fowler, p. 385)

Ever since Descartes, the philosophy of mind within the paradigm of the unicorn-world has been dominated by some pseudo-problems and anomalies such as the interaction between the mind and the brain, levels of existence or of analysis and description, the explanatory gap, emergence, mental causation, supervenience, and reduction. Being used in the construction of Ptolemaic epicycles, these notions generated pseudo-problems. Consequently, philosophers needed to invent other Ptolemaic epicycles, and thus the philosophy of mind has become, over recent decades, a whirlpool of pseudo-concepts and pseudo-approaches for pseudo-problems.

As I have said above, because philosophers in general deal with non-empirical problems, there were some who constructed complicated Ptolemaic epicycles. In general, they strongly defended their approaches. As we already know, the main topic discussed in this book is the mind-body problem.[2] Some interrelated issues created by the mind-body problem – the explanatory gap between neural and

2 „Schopenhauer famously called the mind-body problem a Weltknoten, or 'world-knot', and he was surely right. The problem, however, is not a single problem; it is a cluster of connected problems about the relationship between mind and matter. What these problems are depends on a broader framework of philosophical and scientific assumptions and presumptions within which the questions are posed and possible answers formulated. For the contemporary physicalist, there are two problems that truly make the mind-body problem a Weltknoten, an intractable and perhaps ultimately insoluble puzzle. They concern mental causation and consciousness." (Kim 2005, p. 7)

psychological explanations, levels of existence (ontological levels) vs. levels of analysis or description (epistemological levels), emergence and reduction, the problem of mental causation, supervenience, the problem of representation, and the 'hard problem' of consciousness or human subjectivity or what is it like to be a bat – all have been the subject of many debates. The domino effect seems to join all these problems together. If one of them falls, then all the other problems collapse, too. I will proceed to use the EDWs perspective to analyze levels and reduction vs. emergence (4.1), and qualia in relationship with Kant and the I (4.2).[3]

4.2. Levels and reduction vs. emergence

The notions discussed in this subsection are related to many other notions introduced by various approaches in the philosophy of mind.

Emergence involves certain vertical relationships between low-level and high-level properties. Its history is quite complex because it has many interpretations,[4] and the notion itself became complicated because a particular pseudo-problem (ontological emergence) was mixed up with a real problem (epistemological emergence with strong and weak emergence). Therefore, emergence produced more complicated Ptolemaic epicycles.

When we talk about reduction or emergence, we need to specify what is reduced to what or what emerges from what. We can reduce one property or level to another; or one property or level can emerge from another. Therefore, we are confronted with another complicated notion: levels. In order to explain all these notions, various approaches have appeared in the last century in the philosophy of mind. I will attempt to show the connections between some notions and some approaches, while at the same time analyzing them from an EDWs perspective. (See Chalmers 2003 and van Gulick 2003 for similar classifications.)

(1) On the one hand, we can relate reduction, epistemological levels/ properties and epistemological emergence to identity theory and even to eliminative materialism. The pioneers of identity theory, U. T. Place and Herbert Feigl consider that mental processes or events such as sensations are just physical phenomena: mental processes are nothing but brain processes. Mental processes are, for

3 For mental causation and supervenience, see Vacariu (2008, Chapter 4).

4 Kim wrote „Since around 1990, the idea of emergence has been making a big comeback, from decades of general neglect and disdain on the part of mainstream analytic philosophy." (Kim 2006, p. 547) „Emergence is very much a term of philosophical trade; it can pretty much mean whatever you want to mean…" (Kim 2006, p. 548) Within the unicorn world we can understand this comeback. For a criticism of analytic philosophy from an EDWs perspective, see Vacariu (2007).

instance, the experience of pain, that of seeing something, or that of having a mental image of it.

Ullin Place restricts the identity theory to sensations and mental images. Intentionality, which is a major property of mental states, cannot be constructed as a property of brain structure. (Place 1988, p. 209) Defining the verb is in the sense of strict identity, Smart maintains that sensations are brain processes. „Sensations are nothing over and above brain processes." (Smart 1962/1959, p. 56) Therefore, identity theory reduces the mental level to a neural one.

It seems that the identity theory presupposes a kind of epistemological emergence. Therefore, from an EDWs perspective, it is interesting to analyze Place's distinction between two expressions: the is of definition (as in, „A square is an equilateral rectangle.") and the is of composition (for example, „A cloud is a mass of droplets or other particles in suspension." (Place 1956, p. 34) He argues that the reasoning leading from the logical independence of expression to the ontological independence of entities breaks down when we compare the brain and the consciousness. In fact, „Place spoke of constitution rather than of identity". (Smart 2004, p. 2)

Making an analogy with the mind-brain relationship, Place offers two examples: 'This table is an old packing case' and 'Lightning is an electric discharge'. Being directly interested in such analogies, I will go into more detail regarding Smart's comments on these analogies. Concerning the objection that sensation does not mean the same thing as 'brain process', Place indicates that 'this table' (or lightning, in the second case) does not mean the same as 'this old packing case' (or 'motion of electric charges', in the second case). In different ways, we can distinguish that something is a table (lightning) or an old packing case (motion of electric charges). However, these different ways do not „prevent the table being identical to the old packing case". (Smart 2004, p. 2) Place, Feigl and Smart claim that even if sensation and brain process differ in meaning, they have the same reference. (Smart 2004) Defining is in the sense of strict identity, Smart maintains that „Sensations are nothing over and above brain processes". (Smart 1962/1959, p. 56) According to Heil, this strict identity is applied to processes, events and properties. (Heil 2004, p.79)

Setting aside the EDWs perspective for a moment, I can ask what it means for two terms – for instance, a sensation and its corresponding neural pattern – to refer to the same thing. For a proponent of the identity theory there are, evidently, different conditions of observation for sensation and neural patterns, but these notions refer to the same thing. What does 'the same thing' mean? Is this thing a sensation or a neural pattern of activation? The problem is that if the real thing is merely a neural pattern, then, according to eliminative materialism, we have to reject the second notion, that of sensation. If there is a more fundamental level, we can have a regression ad infinitum.

The same problem has occurred in physics and in the philosophy of physics: at what level do entities really exist? Wittgenstein and all the philosophers who followed him believe that this question has no meaning. We can only ask about the conceptual framework in which we have to include a notion. According to this line of thinking, we can ask: is the mind-body problem a conceptual or a linguistic problem? Do we need to see only how we can conceptually or linguistically define the mind (and its components) and the brain (and its components)? Then, following Wittgenstein (with his investigation of language) and Carnap (1950, with his 'linguistic frameworks'), it seems that, for the proponents of the identity theory, the difference between mental and physical is just a conceptual (linguistic) difference, which reflects conceptual levels or levels of analysis.

In fact, we can notice that the theory of identity provides no ontological solution, but only a conceptual/linguistic solution to the inquiry regarding the existence of physical and mental states. In the Kantian sense, it would mean that one notion, either sensation or belief, or neural pattern of activation, is empty. Which one? Is Rorty right in eliminating the notion of sensation? Are the Churchlands[5] correct in eliminating the notions of folk psychology (consciousness, mind, mental states, self, etc.)? Within the perspective of conceptual frameworks, those who follow Carnap think that we cannot ask about reality, but only about which language/conceptual framework is correct.

In order to avoid transforming the mind-body problem into a linguistic one, we have to answer another question: what do Descartes' internal and external perceptions mean? In a Kantian framework, space and time intuitions, as well as categories, are the transcendental conditions for our phenomenal experience. In the situation mentioned above, can we ask if the observational conditions for sensations/beliefs and neural patterns of activation are the transcendental conditions for phenomenal experiences such as sensations and neural patterns? We return again to the main question: which of these phenomenal things, sensations or neural patterns, really exists? It is impossible for both to exist in the same, real world, because that would lead to a contradiction: we cannot observe (or perceive, in Cartesian terms) two different things in the same place, within the same world, at the same time.

If we accept that these perceptions observe the same thing, only one condition of observation (an internal or external tool) can offer us the real thing. Which of these perceptions or observational conditions plays this role: the internal perception (which would determine the mind) or the external perception (which would determine the brain)? From an EDWs perspective, Places' 'different ways'

5 Patricia and Paul Churchland are the main proponents of eliminative materialism: the mind, the consciousness, and mental states do not exist; these notions are just linguistic games, or erroneous concepts.

involve either organizationally different parts (in the same EW) or epistemologically different entities (in EDWs). Ignoring the distinction between the organizational threshold and the epistemological-ontological threshold, he continues to work within the unicorn-world. Sensation and brain process differ not only in meaning, but also in reference (they are epistemologically different entities) and this is why we cannot reduce mental states to neural patterns. However, a table is either composed of organizationally different parts, or it corresponds to epistemologically different entities. Again, I will emphasize the fact that 'composed of organizationally different parts' is just a theoretical notion introduced by us as observers. The table exists only at the surface, as a whole, when it interacts with other macro-objects.

Even if the position of eliminative materialists is somewhat different, their perspective can be regarded as an offshoot of materialism. The main idea of eliminative materialism is that folk or commonsense psychology is completely wrong. Notions from folk psychology such as belief, sensation, pain, desire, etc., or propositional attitudes and mental causation do not exist. Within the unicorn-world, eliminative materialism has been correct to embrace the Kantian framework of 'empty concepts'.

The Churchlands introduce eliminative materialism from a point of view given by the relationship between folk psychology and neuroscience. We can point out that, officially, they are not reductionists, but eliminativists. If the concepts of folk psychology are wrong, then the statements that employ such concepts are false. The Churchlands eliminate both the mental states and the I from scientific discourse. The Churchlands dedicate an article to intertheoretic reductionism. For them, the reduction takes place from experiential properties of common objects to microphysical properties. (Paul and Patricia Churchland 1990, p. 69) They judge that this reduction is a good thing because the old theory is replaced by the new one. Moreover, the new theory is more simple in explaining nature, it has a much deeper insight and thus has more effective control over and inherits all the evidence of the older theory.

In fact, in the article they offer examples from chemistry and physics to show that (1) the reduction can be domain specific and (2) the reduction of psychology to neuroscience is possible. (P.M. & P.S. Churchland 1990) McGinn points out that Paul Churchland shows that his anti-reductionist opponents confuse ontological issues with epistemological ones. From an ontological point of view, mental states may be identical with brain states and their properties, while from an epistemological point of view, folk psychology and cognitive neuroscience are „two distinct forms of knowledge (knowledge-by-acquaintance vs. knowledge-by-description)" which use two distinct vocabularies (Paul Churchland 1998, p. 156). Due to the continuous progress of science, in the future we will be able to give up the mentalistic description provided by folk psychology. However, according to McCauley, the Churchlands conflate intralevel reductions with interlevel ones.

(McCauley 1996 in Rockwell 2003) Intralevel reduction, which is in fact a kind of eliminativism, refers to successive theories at the same level of analysis; the interlevel thesis refers to the reduction of two theories from different levels of analysis. If intralevel reduction is quite common in science (because one theory can be wrong, in which case it will be replaced by a new one), interlevel reduction (the reduction of psychology to neuroscience) is quite a difficult process. (McCauley 1996, p. 31 in Rockwell 2003)

From my perspective, eliminative materialism seems to be a partly inaccurate alternative because it presupposes the elimination of an epistemological world, the mind-EW which has the same objective reality as the macro-EW where the body (brain) is situated. Thus, the debate between folk psychology and neuroscience is a pseudodebate. If true, a similar argument could be used for the dichotomy of neuroscience vs. the quantum world. Within the realism-antirealism debate, neuroscientific notions can describe entities that do not exist. The point here is that within these disciplines (psychology, neuroscience, and quantum physics) there are different observational tools through which we observe constitutive epistemologically different entities produced by the relationships among them that belong to the EDWs.

In science, objectivity is thought of as being related to spatiotemporal frameworks. In cognitive science it is, of course, unambiguous to study an epistemological world with the external tool of neurology. However, the classical spatiotemporal framework has been made problematic by Einstein's theory of relativity and it has been completely rejected by quantum theory. Interlevel reduction means the reduction of one EW (mind-EW) to another EW (brain-EW), and this is false. Even if the mind-EW with its internal entities (mental states) and mental processes has no spatial framework (it merely has a temporal dimension), this does not mean that it does not have any objective reality. Even if currently the only way to scientifically explain mental structures and functions is to use neural terms, we have to specify that in this case external tools offer us only a loose approximation of correspondences to psychological reality. Using modern technologies, and perhaps by inventing new apparatus in the future, we can come closer and closer to identifying those local firing neural patterns which roughly correspond to mental states and subjectivity, but practically we will not be able, I believe, to scientifically explain the global relation between the most activated neuronal patterns, the entire brain and the body.

Both Patricia and Paul Churchland consider that we will end up eliminating so-called folk-psychology in favour of neuroscience. However, Patricia Churchland maintains that, in the context in which some of a neural network's properties are given by the interactions among neurons and certain rhythmic properties, the notion of emergence has a meaning. Because neural behavior is highly non-linear, the properties of neural networks are complicated functions depending on the properties of parts, not a simple sum of those parts. In this context, high-level properties really exist and we need high-level descriptions to explain them. (Churchland 1996, p. 285) Thus, eliminative materialism asserts that materialism is probably true and that all other alternatives are inadequate to explain human behavior.

Patricia Churchland writes: „The possibility of nontrivial revision and even replacement of existing high-level descriptions by neurobiologically harmonious high-level categories is the crux of what makes eliminative materialism eliminative". (Churchland 1996, p. 286) In fact, she recognizes a kind of weak or diachronic emergence, but not a synchronic or strong emergence. If in the beginning they were complete eliminativists, I think that in recent articles the Churchlands, especially Patricia, have become quite moderate eliminativists. From an EDWs perspective, the emergence accepted by Patricia Churchland involves organizationally different parts, not epistemologically different entities. A specific neural pattern of activation interacts only with patterns with the same properties. The neurons that compose a pattern do not exist for that pattern. Thus, a pattern exists only at the surface.

At the end of this section I introduce a few remarks about two other kinds of reductionism, neutral monism and dual-aspect theory. The main idea of these approaches is that the mental and the physical are reduced to the third level of basic neutral entities. At the end of the 19th Century and the beginning of the 20th Century, neutral monism was one of the first major trends that appeared as a reaction against Cartesian dualism. Neutral monism provides a solution to the mind-body problem, but it also has serious consequences regarding the nature of reality. Trying to go beyond Descartes' problem regarding the interaction between mind and body through the pineal gland, Spinoza considers these two substances to be dual aspects of a third substance, Nature or God.

Neutral monism and dual-aspect theory bring neutral entities or properties into discussion. Spinoza and almost all philosophers after him tried to avoid Descartes' problem: the existence of two different ontological substances within the unicorn-world and the very problematic union between mind and body within the same person. In recent decades there have been certain philosophers who argued for different versions of neutral monism: David Chalmers with his protophenomenal properties or information, and Daniel Stoljar with his o-physical properties (both authors preserving the so-called 'unknown entities') and others that are quite close to physicalism such as Grover Maxwell, Feigl, Michael Lockwood, and Galen Strawson. Diaz includes Russell and Lockwood as supporters of the dual-aspect theory as well. (Diaz 2000, p. 394) Different kinds of entities exist in the world (in Kantian terms they all share the same objective reality), but entities like the mental and the physical are reducible to the basic neutral entities.

(2) On the other hand, we can relate strong emergence and ontological levels to Cartesian dualism and weak emergence and epistemological levels to neutral

monism, dual aspect theory and non-reductionist materialism. As we saw above, believing that the mind and the brain (body) are two different ontological substances, Descartes was aware of an anomaly that was, for him, the interaction or the union between these substances. This anomaly could not be solved within the unicorn-world.

In order to avoid Cartesian dualism, but at the same time to preserve the existence of both the mental and the neural within the unicorn-world, some philosophers constructed levels (a Ptolemaic epicycle which replaced the Cartesian notion of substances) and ontological emergence (namely, properties or entities that belong to a higher-level emerge from a lower-level). The problem is that if we accept ontologically different levels and radical[6] or strong[7] emergence, we arrive at Cartesian dualism, which is rejected by the majority of philosophers.

Nevertheless, to explain some of the phenomena which really exist some philosophers (mainly followers of nonreductive materialism) knew that if we reject Cartesian dualism some anomalies appear. Therefore, some of them consider that human beings are limited entities who can only ever have limited knowledge (McGinn), some take refuge in the distinction between ontology and epistemology (with levels of analysis and the corresponding epistemic emergence, weak and strong), and others introduce the notion of 'organizational levels', which is related to a layered view of nature.

Within the non-reductive physicalist approach, philosophers such as Davidson, Fodor, Jackson, McGinn, Nagel, Putnam, Searle, Chalmers, and Van Gulick have argued in different ways for the epistemological irreducibility of mental qualitative phenomena or consciousness to physical states. In the context of non-reductive physicalism, one problematic notion is qualia, or subjective phenomena. Even if the mind and consciousness are the result of a series of neural phenomena, an explanation of mental phenomena – or at least of some aspects of them like their qualia (qualitative content) or consciousness – cannot be given in neural terms.

6 Radical emergence: „The whole has features that are both 1) Different in kind from those had by its parts and 2) Of a kind whose nature and existence is not necessitated by the features of its parts, their mode of combination and the law-like regularities governing the features of its parts." (van Gulick 2001, p. 17)

7 For Chalmers, a strong emergent property is one in which „the high-level phenomenon arises from a low-level domain, but truths concerning that phenomenon are not deducible even in principle from truths in the low-level domain". (Chalmers 2006, p. 244) If the strong emergent phenomena cannot be deducible from the laws of physics, for instance, then we need new laws of nature to explain such phenomena. The only known phenomenon that cannot be explained by physical laws is consciousness. The examples supporting this idea are those of the colorblind scientist and of zombies. However, consciousness is correlated with or supervenes physical, neural states. „We can think of strongly emergent phenomena as being systematically determined by low-level facts without being deducible from those facts." (Chalmers 2006, p. 247)

Describing the mental qualitative states in causal terms leaves out the special problem of qualia and of consciousness in general. The core idea of irreducibility is that qualia and consciousness are subjective phenomena. Nagel insists that the methods of objective physical understanding „can be used on the body, including its central nervous system", but for the explanation of qualitative phenomena, a „different form of understanding" should be considered (Nagel 1993, p. 66). In Chalmers' interpretation, the phenomenal properties (or qualia) describe the state of a person or an organism in a phenomenal state; these properties belong to the individual, not simply to the mental states themselves. (Chalmers 2003, p. 3)

As far as I am concerned, the meaning of non-reductive physicalism remains very unclear. We cannot explain qualia or consciousness in physical terms, but only from the first-person ontology. But does it mean that qualia and consciousness are still physical elements? Does it mean that they belong to another physical level? Is this non-reductive physicalism an epistemological position about a kind of ontological noumena? (We have to remember that Chalmers adopts a kind of Spinozist monism.) Then, if we use two epistemological notions to describe one ontological element, one notion is empty (in the Kantian sense). Two notions that describe the same entity cannot have the same value, so what values do these notions offer us? Non-reductive materialism is different from the EDWs perspective in some essential points. By looking at its name, we understand that non-reductive materialism refers only to an epistemological non-reduction. For me, there is a hyperontological non-reduction, i.e., mental states exist just as much as physical states do, but in EDWs. However, the main difference is that the mental states are the I, which is, from this point of view, an EW. But because of its unity, the I is, at the same time, an indivisible entity. (For a discussion of qualia and the I from my perspective, see the next section.)

According to van Gulick, emergence can be described in the following way: 'Xs are more than just Ys' and 'Xs are something over and above Ys'. He divides emergence in classes, which contain properties and causal powers or forces. (van Gulick 2001) The emergent features go beyond the features of parts, from which they emerge (p. 16), and this can be categorized as 'metaphysical emergence' (if it refers to the relationship between real things) or as 'epistemic emergence' (if it refers to cognitive explanatory relations about real world items). What does 'over and above' or 'emergent features go beyond the features of parts' mean? Earlier than van Gulick, there were other people insisting upon the difference between ontological and epistemological emergence, considering that people many times conflate them. (Silberstein and McGeever 1999; O'Connor and Wong 2005) Within the unicorn-world, such conflations were inevitable.

Emergence has the same properties as non-reductive physicalism: distinctness and dependence. (Crane 2001) It can involve the whole-part relationship in two different ways. Some properties are merely combinations of parts which exist on the same level, as it were. (For example, a weight of 10 kilos is the sum of 5 + 5 kilos.) Ernst Nagel refers to these as properties which can be treated from an „additive point of view". (Nagel 1963 in Crane 2001, p. 7) Emergent properties are different from the sum of their parts; they imply novelty, which can be quite difficult to define.

In Crane's definition, the novel properties of an object would be „those determinable properties whose determinates are not had by all of the object's parts". (p. 8) From this perspective, surface colour and wetness are novel properties. However, in order to distinguish emergent properties from reductive properties, novelty is not enough. According to Crane, a stronger notion of emergence implies considering it a property of a whole, which has powers that are not related to the powers of its parts. This strong notion of emergence denies that the emergent properties of a whole are supervenient upon the properties of its parts. A dependent property implies that the properties of the whole supervene on the properties of its parts. Therefore, for Crane emergence is strongly related to the supervenience of a whole's properties on the properties of its parts. (Crane 2001, p. 9)

However, for an emergentist, such properties would be 'over and above' the physical properties. From my point of view, we can finally explain the meaning of the spooky expressions 'over and above' or 'emergent features go beyond the features of parts'. There are two ways of explaining the expression 'over and above'. Firstly, if there are two epistemologically different entities, in this case 'over and above' means having two EDWs. Secondly, from an it point of view, the properties of the it are 'over and above' the properties of the organizationally different components of that it. For instance, in the context of a table or a bacterium interacting with their environment, their organizationally different components do not exist. Both these entities and their organizationally different parts exist at the same time only for us as observers. The same observation is valid for surface colour and wetness.

The status of epistemic emergence (weak and strong) is quite complex. For Silberstein and McGeever, Stephan, van Gulick, O'Connor and Wong, Chalmers, and Kim, epistemic emergence is related to our incapacity to explain and predict this type of property of a whole system in terms of its parts. However, the property of the whole is determined by the properties of the parts. For van Gulick, epistemic emergence is subjective and there are weak and strong emergences that characterize predictive and explanatory emergence, as well as representational emergence. For Silberstein and McGeever, we can analyze epistemological emergence only from the point of view given by the relationship between human knowledge and simple or complex systems. This concerns the functional or descriptive analysis of emergence. „Epistemologically emergent properties are novel only at a level of description." (Silberstein and McGeever 1999, p. 186) For Chalmers, a high-level phenomenon represents a weak emergence in relation to a low-level domain when that phenomenon is unexpected according to the laws or principles from the low-level area. Here, unexpected means that emergent properties are somehow deductible from low-level properties (with initial conditions). Examples of such emergence are 'the game of life', connectionist networks, evolution (for intelligent creatures) and high-level patterns in cellular automata. Chalmers introduces a better definition of weak emergence as the phenomenon „wherein complex, interesting high-level function is produced as a result of combining simple low-level mechanisms in simple ways". In this sense weak emergence is a „something for nothing" phenomenon. (Chalmers 2006)

From an EDWs perspective, epistemological emergence reflects the relationship between an entity and its organizationally different parts. As I mentioned above, applying the principle of complementarity, the organizationally different parts and the entity exist at the same time only at the level of description. More precisely, a table or a bacterium does not observe its parts. However, there is an important difference between applying levels of description to non-living and living entities. A table and its legs are simply various levels of description. The sum is equal to the whole. However, this principle is not valid for complex objects (like a car) or for living entities. A car does not exist for its parts (engine, doors, windows, etc.), the parts do not exist for the whole: the car has some properties, but the parts do not have these properties.

Within the same framework, from an EDWs perspective, I want to offer a more detailed explanation of the difference between organizational-analysis levels and their relationship with weak and strong emergence. From an external point of view, an organism and its macro-parts do not belong to EDWs, but only to the macro-EW. For us, organizationally different parts, such as cells, macromolecular, organs, etc., belong to the same EW macroscopic world. One reason is that all these types of entities exist in the same spatio-temporal framework and we use the same observational tools to examine them. Another reason is that all these elements follow the same physical laws.

It would be completely wrong to apply this notion of organizationally different parts to the mind-brain relationship. As with the emergence problem, the explanatory gap between mental and neural levels becomes the relation between two EDWs. The internal tools do not monitor any spatiotemporal feature, and thus it is almost impossible to observe the correspondence between a mental state and certain areas of brain. According to the 6 Principle, mental states involve subjective experience and thus they somehow correspond not only to the most activated neural firing patterns, but also to the rest of the brain and body which continuously interact with the environment.

Consequently, the difficulty of bridging the explanatory gap means that it is practically impossible to identify the most and less activated neuronal pattern that corresponds to each mental state which involves the subjectivity of a person. In general, different concepts refer to entities that belong to either EDWs or organizationally different parts. The entities from EDWs exist because of the epistemological constitutive interactions among each set of epistemological entities that belong to EDWs. At the same time, each observational condition is constitutive in the sense that such conditions reveal the epistemologically different entities to us. Mental representation and neural patterns of activation are not the same entity described at different levels, or organizationally different parts. They are epistemologically different entities that belong to EDWs. The organizationally different parts follow the same epistemological interactions (epistemological laws), while EDWs follow epistemologically different interactions (epistemologically different laws).

4.3. Qualia, Kant and the I

Qualia is one of the most controversial notions in the philosophy of mind. Qualia or qualitative experience involves the I, or human subjectivity, and partial consciousness. According to the principle of knowledge, qualia are (parts of) the I. Consciousness is a part of the I, but it is not identical with it. Conscious and unconscious states and phenomena are the I. For Chalmers, consciousness has a phenomenal structure without any spatiotemporal properties. (Chalmers 2003, p. 26) In his view, phenomenal properties (qualia) – or what it is like to be in a particular state – are not intrinsic properties (otherwise qualia would have to be a substance), being instead related to their definitions. Moreover, from the point of view of the I in the EDWs perspective, I used one of Chalmers's ideas, namely that „Phenomenal properties can also be taken to be properties of individuals (e.g., people) rather than mental states, characterizing aspects of what it is like to be them at a given time". (Chalmers 2003, p. 3, footnote 2) This point of view is quite close to the definition of the I within the principle of knowledge. However, the individual is the same thing as the I. Following Chalmers, Kim is aware of the fact that the problem of qualia and qualitative mental states cannot be completely solved and we have to accept a residue. (Kim 2005)

According to the EDWs perspective, human subjectivity or the I corresponds to the whole brain and body. According to the 9th Principle, all knowledge is the I. How can we reinterpret the Kantian self within this new framework? From a Cartesian framework, or even from an actual scientific framework, the existence of something can be proved only through a clear, distinct and complete perception that produces an explicit knowledge. Within this framework, Kant knew that we could not prove the existence of the I. He was right to reject Descartes' argument concerning the proof of the existence of the I („I think, therefore I am"). Descartes claims that the I exists because it can think and the process of thinking is its essence. (AT VII 78 [CSM II 54] in Wilson 1998, p. 189) As we saw in 1.1 (Principles, I, 9, AT VII 7; CSM I, p. 195), perception is included in the process of thinking. The role of perception involves the connection between thinking and existing. (Wahl 1998, p. 185)

As we saw in 2.8, for Kant the 'I think' does not exist per se but „only for thinking in general; i.e. its reality is the reality of the act of thinking in itself (of judgment)". (Waxman, pp. 832–3) However, even Kant (and other philosophers) considered that thinking can be merely a process or a mechanism and that Descartes did not necessarily prove a relationship between the I and the process of thinking. Therefore, thinking does not prove the existence of the I. The I is, at the same time, indivisible. Kant claims that we cannot prove the existence of the I.

Within an EDWs perspective, his 'bare consciousness', i.e., the consciousness without any qualities, is in fact an implicit knowledge that is different from explicit knowledge. For Kant it is the unity of consciousness or the transcendental apperception that „precedes all data of intuitions… This pure and original unchangeable consciousness I shall name transcendental apperception". (A107) Nagel, Chalmers and Kim are right in considering that a thing is functional when there are tools of observation for it.

Human subjectivity and qualia are not functional. From an EDWs view, the I is not functional, because the I has no tool to observe itself as a whole. In Kantian terms, we do not have any determination or representation for the I. Do all the above elements completely deny the existence of the I? No, because otherwise I could ask myself who is writing this book. Is it only a brain and some separate elements? Maybe somebody would answer that the only things involved are some mechanisms for perception, for thought, for moving the fingers and the head and so on, which are all encapsulated within a body, and that their functions are coordinated as the result of the evolution of species.

Kant affirms that „[w]e are conscious a priori of the complete identity of the self in respect of all representations which can ever belong to our knowledge, as being a necessary condition of the possibility of all representations". (A116) We are not conscious a priori of the complete identity of the self. In fact, the I involves conscious and unconscious (and/or explicit and implicit, controlled and automatic) elements. The entire body (that includes the brain) corresponds to the I. The neural patterns of activation of the brain have different degrees of activation. Some of them correspond to conscious elements, some of them to the unconscious part of the I. The brain is so strongly related to the rest of the body that we cannot avoid this relation in analyzing the I that corresponds to the brain-body relation. This unity represents, in the Kantian sense, the unity of knowledge, or more exactly implicit knowledge.

In a short paper, a famous neuroscientist, Raichle hints at a special topic in neuroscience: the dark energy of the brain. The question is: what does the brain need so much energy for? „The brain apparently uses most of its energy for functions unaccounted for – dark energy, in astronomical terms." (Raichle 2006, p. 1249) In modern times, using PET and fMRI, researchers realized that the energy necessary for the brain to manage the demands of the environment is less than 1% than the actual quantity of energy used. The brain's metabolism and its circulation require only a little of the energy consumed by the brain. In this context, the logical answer seems to be that the energy is necessary for the intrinsic activity of the brain. But what does 'intrinsic activity' mean?

Raichle analyzes some possible answers to this question:

a) Spontaneous cognition (our daydreams, or the stimulus of independent thoughts. However, his comment on this possibility is that the brain responds with a small amount of energy to controlled stimuli, so the energy cannot be larger for the stimulus of independent thoughts)
b) Facilitated responses to stimuli (while receiving continuous excitatory and inhibitory input, neurons - patterns of neurons and large parts of the cortex - pass through various balances that determine their responses)
c) Interpreting, responding to and predicting environmental demands.[8]

Finally, Raichle suggests that further research is needed to clarify the spontaneous activity of neurons. (Raichle 2006, p. 1250) My position, from the EDWs perspective, is that dark energy represents the relationship between the large parts of the brain and the body. The dark energy of the brain (which includes Crick and Koch's notion of penumbra), the entire brain and the body correspond to the I. I regard this dark energy as representing the relationships between the large parts of the brain, the body and the environment. We have to take into account that, after a period of training, neural patterns are less activated in accomplishing a task.

I believe that dark energy (which reflects the relationship between the large parts of the brain and the body) corresponds to implicit, procedural unconsciousness and automatic knowledge and is the result of the evolution of species. It corresponds, in fact, to the Kantian empty or bare consciousness: 'I think'. That which thinks can be represented only as something 'purely intellectual', as an empty thought. (B157) In this case, there is only an indeterminate given object ('indeterminate perception' or 'bare consciousness' (A 346/B404)) which is incorporated in that act of thought, therefore we cannot apply the categories. The 'I that thinks' is just a 'bare consciousness' that cannot completely prove its existence.

8 For more details about this „dark energy of the brain" or, as it has been called more recently, the „default mode network", see Vacariu (2014, Chapter 5).

It is the possibility of our cognition (spontaneous thoughts or the stimulus of independent thoughts) and our experience (interpretations, responses and predictions relating to environmental demands). This is the reason that, using fMRI and PET, we cannot observe/understand 80% of the brain's energy or Raichle's dark energy.

Related to the I, Kant could not explain the spontaneity of thoughts. From an EDWs perspective, there are two ways to explain the spontaneity of thoughts. Firstly, as explicit knowledge, a spontaneous thought is the result of the reactions of implicit (unconscious, etc.) knowledge. Nevertheless, it is very difficult to explain in detail why a person has a spontaneous thought at a certain moment. Secondly, we can try to grasp which biological elements correspond to a spontaneous thought.

Raichle considers that the explanation of the spontaneous activity of neurons is one of the current major tasks of neuroscientists. (Raichle 2006, p. 1250) Evidently, a spontaneous thought would correspond, with rough approximation, to some observable neural interactions (which correspond to explicit knowledge) as a result of some unobservable neural interactions (which correspond to the implicit knowledge that is the empty I). These spontaneous neural patterns of activation are the results of the silent part of the brain (these represent, within the EDWs perspective, some active parts of the brain which cannot be grasped). As is well known, after a period of training certain elements of explicit knowledge become implicit knowledge. But both of these types of knowledge are the I.

During the first steps of training for a task, a large surface of the subject's cortex is involved in solving that task. After these steps, the surface involved becomes smaller. (Baars 1988) It seems that, in some cases, habituation is exactly this process whereby explicit knowledge becomes implicit, which is a result of the species' evolution. Moreover, we have to be aware that habituation takes place for every type of task we perform, even if some of these processes are the results of the evolution of our species, while others are not (e.g. walking vs. riding a bike).

The conscious and unconscious mental states (explicit and implicit knowledge) and various internal feelings in parallel are all the I, that has its unity. More exactly, this unity is the I. Without this unity, there is no I, and Hume would be right in claiming that the I does not exist, but is only an aggregate of various perceptions and ideas. It is difficult to identify which particular parts of the brain correspond to certain mental states that are the I. We can make a parallel analogy between two pairs: electrons-table and neurons-the I (the mind). Without being able to get out of the micro-EW, a micro-being could not ask what the unity of a net of micro-particles is (for example, in the case of particles corresponding to a table from the macro-EW). It is meaningless to ask about the unity (in Cartesian terms, about the extension) of a table from the point of view of electrons. Also it is meaningless to ask about the unity of a table from the point of view of a table leg or about the unity of a bacterium from the point of view of one of its proteins. In the same way, we can only find the correspondences between mental representations and neural patterns. We can construct judgments that have objective validity only if we do not mix terms from the mind-EW with terms that refer to the brain (body) which belongs to the macro-EW (i.e., mixture terms referring to mental states and neural states). We can construct a concept such as 'A certain mental state corresponds, with approximation, to some neural patterns of activation', but we cannot construct judgments that include, for instance, the emergence of the mental from the neural or the mental causation of the body's movement.

I will end this section with Waxman's words about Kant's philosophy. The role of understanding in representing the world for Kant is this one:

> Thus do the categories become effectively the template of the sensible universe. More strikingly still: the understanding, in implementing this Bauplan by means of imagination, and thereby extending the scope of consciousness (that is, of that which is something for me), is actually doing nothing more than furnishing the I-concept with an expanded instantiation. The outcome of Kant's theory of understanding could therefore be expressed as follows: the world is not simply my world, as with other subjective idealist philosophers; the world, for Kant, actually is the self. (Waxman, 1995, p. 857)

Obviously, this refers to our mental images of the external world (external to the brain/body and not to the mind) that are the self. (Fig. 6) The external world is 'inserted/brought' inside the I. However, from the EDWs perspective, the mind does not exist for the external world since an EW does not exist for any other EDW. So, the mind-EW does not exist for the macro-EW, and the macro-EW does not exist for the mind-EW. It is only the brain and the rest of the body that interact with the external environment, but the brain/body and the external environment of the body are parts of the macro-EW, and they correspond to the I.

5
EDWs and Cognitive Science

As we have seen, Chapter 4 is dedicated to the relationship between the EDWs perspective and several notions from the philosophy of mind. In this chapter, I will investigate the relationship between my EDWs approach and cognitive science, a field of study which emerged during the mid-twentieth century, and which unites people working in neuroscience, psychology, linguistics, computer science, artificial intelligence, robotics, anthropology, and philosophy.

The main goal of cognitive science is to find a viable solution to the mind-brain problem. After more than half a century of scientific research, it still hasn't been found. In the 70's, a sub-domain of cognitive science was created, and it was named cognitive neuroscience. (I will discuss it in the next chapter of this book) In this chapter, I will examine the main ideas of computationalism and the dynamical system approach in relation to the EDWs perspective.[1]

Then, from the EDWs perspective, I will analyze some of the key elements of cognitive science (representation, levels of analysis, primitives, processes, structures, threshold, self-organization, bidirectionality, emergence, habituation, and tasks) which partly involve these approaches. Then I will attempt to establish the relationship between these key elements and several philosophical distinctions (continuous-discontinuous, as well as static-in motion).

5.1. Computationalism

In this section I will present certain elements of computationalism (I am referring to the Computational Theory of Mind, hereafter CTM). CTM was initiated by Putnam (Putnam 1961), but its most important contemporary proponent is Fodor (Fodor 1975), followed by Fodor's friend and collaborator Pylyshyn. (Fodor and Pylyshyn 1988) CTM considers that the mind works like an information processing system which computes mental representations, and this naturally implies the use of the Representational Theory of Mind (RTM).

1 To read more about the EDWs perspective, connectionism and robotics, see Vacariu 2008, Chapter 5.

Until recently, the generally accepted paradigm for CTM regarded behavior as the outcome of computations performed over representations in the mind. Within this classical paradigm, cognition essentially involves representations and computations. The main difference between classical models and other theories (which similarly assume the existence of representations) is that the former consider that mental representations are built according to combinatorial semantics and syntax. This means, structurally speaking, that representations have syntactic constituents, and the semantic content of each representation is a function of the semantic content of syntactical parts that belong to the constituent structure of that representation. These representations are complex symbolic structures built upon combinatorial syntax and semantics, and computations are rules that allow the manipulation of proper symbols (Fodor and Pylyshyn 1988). These symbols constitute the Language of Thought (LOT).

One important aspect that needs to be mentioned here is that the CTM offers a good explanation for propositional attitudes (mental states regarding a proposition) which involve judgments. The Language of Thought Hypothesis (LOTH) appeared in the 1970s when propositional attitudes such as hopes, beliefs, desires, etc., were a common topic for the philosophers of mind. The LOTH concerns the nature of thought, in terms of propositional content, without having an answer for problems regarding qualia, phenomenal experience, mental images, sensory memory, hallucinations, etc. (Aydede 2004) Two things inspired Fodor: (1) The notions of formality and computation, syntax and semantics; and (2) The concepts of compositionality, systematicity, and productivity in LOTH.

(1) Formality and computation, syntax and semantics

Euclidian a priori intuitions are essential elements in Kant's theory. In the 19th century non-Euclidian geometries appeared, which were formally consistent without implying any form of empirical intuition. According to Horst, because of the crises produced by the non-Euclidian geometries in the late 19th and early 20th centuries, mathematicians tried to eliminate the appeal to such things as intuition, and developed the concept of 'formal symbol manipulation and computation'. First Bolzano (who directly rejects the Kantian notion of an a priori spatial intuition) and then Gauss, Peano, Frege and Hilbert tried to replace intuitions with formal calculus in geometry. „The most influential strategy for formalization was that of Hilbert, who treated formalized reasoning as a symbol game, in which the rules of derivation were expressed in terms of the syntactic (or perhaps better, non-semantic) properties of the symbols employed." (Horst 2007, p. 2)

However, Clark mentions the pioneers of formal logic from the 17th century: Pascal and Leibniz, followed by Boole, Frege, Russell, Whitehead, etc. Formal logic is a system that has a set of symbols, the possibility of combining them, and rules for such combinations. (Clark 2001, p. 9) Within this system, the truth of sentences can be preserved by following certain rules. It is amazing that formal logic can preserve „at least one kind of semantic (…) property without relying on anyone's actually appreciating the meanings (if any) of the symbol strings involved". (Clark, p. 9) However, Clark mentions that meaning is „in a certain sense recreated, in a realm whose operating procedures do not rely on meanings at all". The symbols (tokens) are manipulated according to certain rules that act on their physical or syntactic characteristics.

Following Newell and Simon with their physical-symbol systems, Haugeland considers that „If you take care of the syntax, the semantics will take care of itself". (Haugeland 1981a, p. 23, original emphasis in Clark 2001, p. 9) The essential thing here is not the physical part, but the computations (based on certain rules) realized by the system in manipulating the symbols. The idea that the mind is equivalent to software was the framework that allowed Newell and Simon to elaborate the 'physical-symbol system' hypothesis. Their idea is that „[a] physical symbol system has the necessary and sufficient means for general intelligent action. (Newell and Simon 1976, p. 87)" (Clark, p. 28) This framework sends us directly to the distinction that the mind (high-level) and the brain (low-level) are different levels of description or analysis. According to Clark, the high-level (psychological/conceptual level) (symbols + manipulation of symbols = computation) has to be formed of „semantically transparent systems" (qtd. in Clark 2001, p. 29)

From an EDWs perspective, we can see that within the unicorn-world researchers have been forced to find the direct relationship between the 'physical symbol system' and intelligent computation. In my approach, the physical system and the computation are not within the same EW, belonging instead to two EDWs. Obviously, within the physical system there has to be something that corresponds to computation and representation, but it is 'impossible for us to use' these notions within the brain-EW. An essential point for the CTM is the causal role of representation: in the cognitive economy of an intelligent system, the causal role of representation ensures the systematic correspondence of the system to the situation that it represents. The causal role of a complex representation is directly given by the causal effects of its constituents, which are causally and systematically adequate to their content. Thus, the classical explanation of cognitive processes depends upon the causal role of constituents and guarantees that these constituents have a causal role: the complex representations are directly constructed from the instances of their constituents. This simply means that symbolic representations are syntactically structured.

However, CTM has serious problems when it comes to explaining the relationship between syntax, semantics and causation. Any sentence in any particular language is composed of several words. These words and their relationships (certain syntactic rules) represent the syntax. The meaning of the sentence is generally furnished by the meaning of the words which form it. However, in many cases, the meaning of a word is furnished by its context. In order for something (a human being or a computer, for instance) to produce a sentence, it is necessary for it to have a mechanism which acts on (causes) syntactic structure. One of the mechanisms of a computer, for instance, can cause some changes to the syntactic structure of programs. Meaning does not exist within the computer.

The relation between causation and syntax is clear, because syntax represents the formal features of a 'well-formed formula'. However, it becomes difficult to explain this relation once semantics are introduced. Judgments involve not only the syntactic relations among constituents (words), but also the meaning of those constituents (semantics) and the formulae in which they occur. From an EDWs perspective, both mental representations and their semantics belong to the I. According to Horst, syntax is the intermediary between semantics and causation. (Horst 2007) For such things, the notions of formalization and computation are necessary. Therefore: „Formalization shows us how to link semantics to syntax, and computation shows us how to link syntax to causal mechanisms". (Horst 2007)

But in what sense can formalization and computation work with syntax and semantics? Syntax, semantics and causation correspond to neural processes that take place in the brain, but which we cannot identify. A computer manipulates symbols without understanding the meaning of the sentences/formulae that are constituted by words/symbols. Such a device is called by Haugeland a 'semantic engine'. (Heil 2004, p. 108) The problem is as follows: on the one hand, the computer does not understand the meaning of symbols, even if it displays words and sentences. Even if those meanings are meant for us as external observers, the syntax, which the computer manipulated, is considered to be the mirror of semantics. (Heil 2004, p. 111) On the other hand, the I manipulates symbols, but at the same time has access to their meanings.

Can the I or the mind-EW be similar to a computer? This problem is strongly related to the Chinese Room Argument (which is discussed below) and to connectionism (see Vacariu 2008, Chapter 5). The mind-brain relation can be likened to the relationship between software and hardware in the case of a computer. If an external observer – using dissection, or fMRI and PET – looks inside a brain, she can see nothing that resembles words and sentences. The same holds true for what happens inside a computer: on one level, electrical changes happen to different elements and, on another level, it runs a program. However, the confusion appeared when everybody became interested in finding which parts of the brain are similar to the central processing unit of a computer. According to the EDWs perspective, all the knowledge is I that corresponds to the union between the brain and the body. There is no central processor in the mind.

There are several objections against this analogy between the mind and a computer. I will specify only the most important one: Searle's Chinese Room. With his thought experiment, Searle shows that a computer, following syntactic rules for manipulating strings of symbols, does not understand anything about the symbols being manipulated. In fact, Searle's argument is against the CTM, too. In the article from 1984, his argument against Strong AI[2] had the following demonstration:

(1) A computer has a formal (syntactic) program.
(2) Syntax is not sufficient for semantics.
(3) The human mind has mental content (semantics).
Therefore,
(c) Programs are not sufficient for the mind.

This argument shows that formal systems are not sufficient for mental content. It implies that for computers symbols have no meaning, while for the human mind they do. (Searle 1999) According to Searle, the notion of understanding is misunderstood. He believes that there are different degrees of understanding, for example, being able to understand stories in English or in French well, understanding them less well in German, and not at all in Chinese. Moreover, there are different kinds or levels of understanding. He points out that we use metaphor and analogy to attribute understanding to human artifacts such as cars and thermostats, but 'nothing is proved by such attributions.' (Searle 1980) This is the essential difference between what understanding means for us and what it means for a machine. The car, the thermostat, and the computer all understand in the same way, i.e., they understand nothing. In this way, Searle rejects Strong AI.

(2) Compositionality, systematicity, and productivity in the 'language of thought hypothesis'

Making an analogy between language and thought and thus forwarding Chomsky's theory about grammar[3], Fodor elaborated the 'language of thought hypothesis', which attributes certain features to the process of thinking: compositionality, systematicity and productivity. Fodor and Pylyshyn have strong arguments in favor of these three attributes. (Fodor and Pylyshyn 1988)

Compositionality refers to the fact that the content of a complex representation is determined by the content of its constituents and their relationships.

2 „But according to strong AI, the computer is not merely a tool in the study of the mind; rather, the appropriately programmed computer really is a mind, in the sense that computers given the right programs can be literally said to understand and have other cognitive states." (Searle 1980, p. 417)

3 Contrary to Skinner's behaviorism (in every child's case, the entire language is simply acquired during the first years of development), Chomsky considers that language is innate for humans. More precisely, some rules (the universal grammar which exists at the core of all languages) and the nature of representations/words are innate.

Systematicity is best represented by the following example: if someone understands the sentence 'John loves Mary' she can also understand, without additional learning, the meaning of 'Mary loves John'. Productivity implies that the human mind is capable of using recursive syntactic rules and a finite set of lexical representations to produce a virtually infinite number of thoughts.

From an EDWs perspective, the I is able to transform one sentence into another only because both sentences are parts of the I. As we saw above, syntax and semantics are both the I, and this is the reason why the I (the 'thinking I') has the properties of compositionality, systematicity and productivity.

5.2. The dynamical system approach

More recently, some authors have claimed that the dynamic systems theory is the most appropriate framework for understanding cognition. Cognitive systems are considered to be dynamical systems, as van Gelder suggests: „cognitive agents are dynamical systems and can be scientifically understood as such". (van Gelder 1999) The new metaphor discusses the core notions of the preceding paradigms – notions such as computation and representation. Within this new metaphor, there are already tendencies to classify dynamicists in various ways: representationalists, non-representationalists or meta-representationalists, computationalists and non-computationalists, connectionists and dynamic connectionists, and so on.

A dynamic system is characterized by a set of state variables and a dynamic law that governs the changes in their values over time. The set of all possible values of state variables constitutes the system's state space. The parameters of the system determine the dimensions of its space. Each state of the system is a point in its state space. The sequence of the states represents the trajectory of the system. The behavior of a system (which changes over time) is represented by a sequence of points in its phase space (a numerical space described by differential equations). (van Gelder and Port 1995, p. 5) Usually, geometric images are used to determine this trajectory: „behaviors are thought of in terms of locations, paths, and landscapes in the phase space of the system". (van Gelder and Port 1995, p. 14) The main notions are control parameters (factors that affect the evolution of a system) and collective variables. A common example of a dynamical system is the solar system in which the position and the momentum of one planet differ from that of other planets and mathematical laws relate the changes over time. (van Gelder 1995, p. 363 or van Gelder and Port 1995) In fact, scientists try to explain this type of real system by means of a mathematical dynamic model. The rates of change are represented by differential equations.

State space is a set of possible trajectories; the conceptual tools of discrete or continuous mathematics are used to describe the laws that give the shape of possible trajectories (the flow). The main concepts used in order to describe state space are the following: (1) an attractor is a point or a region with the property that any trajectory which passes near it is attracted to that point or region; the surface where this influence takes place is called the basin of attraction (2) the repeller is a point or a region which has a property contrary to that of an attractor: it rejects all the trajectories which are passing near it (3) a bifurcation is a point in which a small change in the values of certain parameters can change the direction of the state space's flow and can shape a new space of attractors and repellers.

Radical dynamicists believe that terms such as representation and computation are useless in any explanation of human cognition. In the framework offered by the dynamical systems theory, cognition is viewed in motion. The Cartesian distinction between mind and body is abandoned. The mind, the body and the environment are dynamical coupled systems, which interact continuously, exchanging information and influencing each other. The processes happen in real, continuous time. In connectionism, van Gelder claims, a change in system is a transformation from one representation to another, these being static entities which exist only at an instant of time: „they result from freezing the behavior of the system". (van Gelder 1995) In a dynamical system we do not have discrete identifiable steps in which one representation gets transformed into another one. From this perspective there are two points of view regarding the problem of representation. The radical one considers that the brain does not compute representations; representations do not exist in the brain/mind. (van Gelder 1999; Kelso 1995; Thelen and Smith 1994; Skarda and Freeman 1988) The moderate one suggests that we only need to replace the vehicle of representations, or to understand the notion of representation in a weaker sense, that is, we cannot completely avoid the notion of representation in explaining human cognition. (Bechtel 1998; Clark 1997a, b; Wheeler and Clark 1997)

From an EDWs perspective, there are many errors embraced by the proponents of this approach. Mental representations exist within the mind-EW, and these entities are, at the same time, the I. The mind is not coupled with the body since the mind is an EW and the body (brain) belongs to the macro-EW. Only the brain is coupled with the body and the environment. The introduction of mathematical tools does not help us study the mind, or at least it does not do so now. It seems that some authors want to convince their readers that this approach is correct by appealing to certain complicated mathematical tools, even if it is impossible for them to apply such notions in explaining cognition. Usually, the examples offered by the proponents of the dynamical system theory (DST) are taken from a physical system (the Watt governor, van Gelder 1995) or from sensorimotor control systems – such as learning how to walk (Thelen and Smith 1994) or doing rhythmic finger motions (Kelso 1995).

For radical dynamicists, high cognitive processes are just the result of the evolution of perception and of sensorimotor control systems. Thus, if the latter phenomena are the result of continuous reciprocal causation between the brain, the body and the environment (which Clark names 'on-going coupling'), then cognition is in the same situation. Rejecting the use of representations in explaining cognition, the proponents of DST consider that representations can be replaced by certain dynamical processes. In DST, the rules are defined over numerical states; dynamical systems are representational „without having their rules of evolution defined over representations". (van Gelder and Port 1995, p. 12) For van Gelder, if everything is in motion, then we do not have any static and discrete representations. Thus, he tries to prove that everything is in motion and „everything is simultaneously affecting everything else". (van Gelder and Port 1995, p. 23)

Dynamicists conceptualize cognitive processes in geometric terms. The distinctive character of a cognitive process as it unfolds over time is a matter of how the total number of states that the system passes through are spatially located with respect to one another and to the dynamical landscape of the system. (van Gelder and Port 1995, p. 15) Cognitive processes always unfold in real time; their behaviors are pervaded by both continuities and discretenesses; they are composed of multiple subsystems which are simultaneously active and interacting; their distinctive kinds of structure and complexity are not present from the very first moment, but emerge over time; cognitive processes operate over many time scales, and events from different time scales interact; and they are embedded in a real body and environment. (van Gelder and Port 1995, p. 18) In their view, the embeddedness of cognition within the neural system is the framework for the idea that a system can be simultaneously described on two levels: the dynamical level and the computational level.

In fact, the dynamics of central cognitive processes and the dynamics of neural processes are two levels (high and low) of description. „Dynamical systems theory provides a framework for understanding these level relationships and the emergence of macroscopic order and complexity from microscopic behavior." (Van Gelder and Port 1995, p. 29) However, the proponents of this approach try to force us to accept that discreteness (mental representations) can be represented through continuous processes. This is clearly false: within the mind-EW, we have static and discrete entities which cannot be replaced by continuous processes. The distinction between appearance (discrete) and reality (continuous) is false; it reminds us of the Kantian noumena-phenomena distinction.

5.3. Dichotomies concerning the notion of mental representation and processing

There are other approaches, more or less along the same lines, which come from different domains and have different proponents: Skarda and Freeman (1987) in Neuroscience; Port and Van Gelder (1995), Van Gelder (1995) in Philosophy; Brooks (1991), Beer (1995) in AI; Kelso (1995), Varela et al. (1991), Globus (1992, 1995), etc. These approaches are rooted in the work of J. J. Gibson, Berthalanfy, Waddington, and others. By studying them, we notice that there are various interpretations of the term representation, which can be situated, for a more suggestive illustration, on an imaginary axis: on the extreme left we find Chomsky, Fodor, Pylyshyn and others, who consider that cognition is based on computations over symbolic representations, and on the extreme right we can situate the perspective influenced by Artificial Intelligence, such as Brooks (1991), Beer (1995), Harvey (1992), and so forth, or Neuroscience, such as Skarda and Freeman (1987), who assume that cognitive systems do not use any kind of representation. Authors like Bechtel, Clark and Wheeler claim that even in the case of a dynamical system, we are dealing with very low-level representation. (Wheeler and Clark 1999; Bechtel 1998; Clark 1997a, b) In cognitive science there are several essential dichotomies concerning the notion of representation. From an EDWs perspective, this imaginary axis is the result of a mixture between the brain-EW (which belongs to macro-EW) and the mind-EW (the I).

Mandler synthesizes these dichotomies in pairs: declarative-procedural, accessible-inaccessible, conscious-unconscious, conceptual-sensorimotor, symbolic-subsymbolic, and explicit-implicit. (Mandler 1998, p. 265) These dichotomies are interconnected and overlap partly, without being identical. (Mandler 1998, p. 265) I will now refer to her summary of procedural and declarative knowledge. Procedural knowledge (perceptual and sensorimotor) has the following characteristics: it is inaccessible to conscious awareness, difficult to describe in language, slow to learn, context-bound, sensitive to frequency, and not represented in rule-like form. The conclusion is that for procedural knowledge a subsymbolic (a connectionist level between computational and neuronal level, Smolensky 1988) format is better, rather than symbolic forms of representation. According to Mandler, declarative (conceptual) knowledge is accessible to consciousness and describable in language, it requires attention to be encoded in this format, it is potentially learnable in a single trial, it has a static character, and symbolic representation. (Mandler 1998, p. 268)

From my point of view, the I cannot access two explicit elements at the same time. However, the I can access a representation of explicit knowledge and several elements of implicit knowledge at the same time. But, as we already know, all these representations are the I. In Mandler's view, symbolic representations proper to a computationalist perspective are best suited to resolving the tasks involved in declarative knowledge, whereas subsymbolic representations (proper to a connectionist approach) are suited to solving the tasks involved in procedural knowledge. The I uses explicit or implicit elements that can be continuous or discrete. Usually, declarative knowledge involves certain static and discrete representations, whereas procedural knowledge involves not static and discrete primitives, but only certain continuous processes.

In this sense, let us consider the example of a subject learning a foreign language or logic. As a novice, she starts with a set of general instructions and relies explicitly on declarative knowledge, not on implicit or procedural knowledge. In order to reach the expert level through training (solving problems and exercises, constructing statements using various new words and rules, etc.), a part of the explicit processing becomes implicit. This happens because of habituation, but here again we focus on the notion of a threshold. This training involves a threshold of habituation that is different from organizational and epistemological thresholds. The knowledge of logic is declarative knowledge, but according to Mandler the process involved in declarative knowledge is either implicit or explicit, while the process necessary for procedural knowledge can only be implicit. (Mandler 1998, p. 267)

Mandler emphasizes that we are aware of perceptual knowledge (or sensorimotor knowledge), but most of this knowledge is not consciously accessible because of its procedural form of storage in memory: „… it can only be run“. (E.g.: face recognition) (p. 266) As opposed to this, verbal information is formed in conceptual or declarative form: recollection, planning, etc. What happens within the brain when a subject passes from novice to expert? In the beginning, in the case of a novice, the surface of activation patterns that corresponds to that task is very large. After being trained in that task, the surface of activation patterns becomes smaller. (Baars 1988) As mentioned above, the indivisibility of the I is a stability which implies conceptual constancy. This means that this stability is maintained, even though through habituation the corresponding surface of activation patterns related to that task is reduced.

Let us consider an example concerning procedural knowledge given by Mandler: tying a shoelace. How do we carry out this action? What elements are involved when we repeat this process? Of course, we appeal to memory (which is the I), but it is not clear what kind of information is stored in our memory. There are two solutions: either we memorize discrete representations, or we memorize certain processes. However, the I continuously moves its virtual arm (that is part of the I and corresponds to a physical arm). What happens in the brain (which belongs to the macro-EW in this case)? There are two alternatives. In the first one, the brain sends a command to the hand's muscles, using discrete representations, and after a very short time it receives feedback from the body regarding what is happening there. The process is repeated many times, so to us it appears to be a continuous process. In the second one, the brain memorizes certain processes, i.e., the representations involved are a kind of representation in motion.

Dynamic system theory rejects the existence of representations and introduces an essential parameter: time. (Port and van Gelder 1995) Usually, dynamicists give examples of bodily actions such as a child's walk (Thelen and Smith 1998) or the movement of fingers (Kelso 1995), and they extrapolate the conclusions from procedural to declarative knowledge. They do not take into account concrete declarative tasks. Thelen and Smith, van Gelder and other dynamicists replace static and discrete representations with attractors that are constantly moving and, on a conceptual level, these attractors seem static and discrete. I claim that the symbolic representations that are the I are static and discrete exactly as a table is static and discrete within the external macro-EW. Within the micro-world, a table does not exist, merely the corresponding network of microparticles (some in motion) exists. The macro-EW that is our ecological niche has certain stable entities.

5.4. The EDWs perspective and some key elements in cognitive science

Within all the approaches of cognitive science there are some key elements by means of which various aspects of human cognitive behavior can be captured. However, these key elements have been constructed within the unicorn-world. In this section I will try to explain the relationships between each key element and EDWs. What I take to be key elements, from which I will start my analysis, are as follows: levels of analysis, primitives, processes, structures, threshold, self-organization, bidirectionality, emergence, habituation, tasks, the interaction between levels, and also the interactions between elements of the cognitive system and the environment. These elements always entail certain philosophical distinctions such as continuity-discontinuity, (state of) motion-(state of) rest, variability-stability, part-whole, and micro-macro. I will show that placing all these key elements within the unicorn-world requires the construction of various complicated Ptolemaic epicycles.

Firstly, I will consider the levels of analysis. As I wrote above, scientists are generally not interested in ontological levels. They analyze what they can observe without asking too much about the nature of observable entities. In cognitive science, researchers consider that there is a cognitive architecture with a hierarchy of levels, but, in order to avoid empty philosophical debates, they simply consider that these are levels of analysis. This context has led to the appearance of a hot question: how many 'levels of analysis' are required to explain cognition? In order to avoid an infinite regression, some researchers think that we must confine ourselves to a limited number of levels of analysis. In this sense, the most familiar positions are Marr's – with three levels of analysis: computational, algorithmic-representational, and implementational – and Smolensky's – again with three levels: conceptual, sub-conceptual, and neural (Smolensky 1988). Sometimes other levels of analysis are also taken into account: either lower levels, such as cellular, molecular, genetic and even quantum levels (Bickle 2008, 2007a, b; Globus 1995; etc.), or higher levels such as the body, the brain, environment connections, situated action, society, and culture.

Other approaches include the traditional reductionism focused on bridge laws between levels, Churchland's eliminativism, Fodor's irreductibility of psychology, and so on. Within the EDWs perspective, we know that concepts from different levels cannot refer to the same entity/event/process because some levels are in fact EDWs. Various concepts refer to entities that belong to EDWs or to organizationally different parts. These entities exist either because of the epistemologically constitutive interactions within each set of entities that belong to the EDWs, or because of the organizationally different interactions that belong to epistemologically different parts. Mental representations and neural patterns of activation are not the same entity described from different levels of description or levels of analysis. They are epistemologically different entities which belong to EDWs and therefore mental representations do not exist for neural activation patterns (and vice-versa). Therefore, talking about levels (it does not matter what kind of levels) is meaningless. For us, each set of observational conditions is, in restricted Kantian terms, constitutive, i.e. only revealing and creating certain entities and processes that belong to EDWs. In fact, the epistemologically different interactions are epistemologically constitutive in creating the epistemologically different entities and processes that follow epistemologically different laws.

As we have seen above, some philosophers are trying to explain the controversial concept of emergence. There is another pseudo-concept in cognitive science related to this one, bidirectionality. The impossibility of explaining cognition on a single level has been acknowledged, thus emphasizing the emergence of primitives from one level to another and the bidirectionality of cognitive processes, usually between neural and conceptual levels (Fischer and Bidell 1998; Black 1991; etc.).

Bidirectionality means that the primitives of the conceptual level emerge from the neural level, and that the conceptual level influences the activity of the neural level. However, this notion of bidirectionality is equivalent to both downward causation and emergence taken together. Black was among the first to insist on this idea with arguments from neuroscience. For other people, this was an essential idea: the continuously circular causation that exists between conceptual and neural levels. Changes that appear at a latter level imply changes at a former level and vice-versa. The idea of bidirectionality determined many people to draw the conclusion that cognition cannot be explained using a single level, be it either conceptual or neural. Therefore people like Gazzaniga and Kosslyn created another special science, cognitive neuroscience. This continuous circular causation reveals our inability to fully explain an EW. Such circular causations are meaningless within the EDWs perspective.

Evidently, the continuous circular causation is a pseudo-notion. As we saw in Lungarella and Sporns' research (2006), there is a continuous circular causation between the brain, the body and the environment. However, according to EDWs perspective, the existence of this sort of EW-transcending causation is impossible. Otherwise, what are the constitutive interactions which form this continuous causation? We can accept emergence or, in our terms, organizational emergence, only between organizationally different parts; that means, for instance, that an organism is composed of its parts—this is merely weak emergence. However, weak emergence is a theoretical notion imposed by us as thinking observers. The organizationally different parts of a table exist only for us as observers, but not for the table itself, or for other macro-objects that interact with (parts of) the table. The table and all epistemologically different entities except the I exist only at their surface in EDWs.

The conceptual level cannot be reduced to the neural level because of the representational character of mental symbols, which is an argument against reductionism and against the unity of (cognitive) science. (Fodor and Pylyshyn 1988) In neural networks with distributed representations these types of 'context-free symbols' do not exist, being replaced by „context-sensitive responses to inputs". (Clark 1997c, p. 180) Even in the case of mental representations, only their surface exists within the I, even if they are the I.

Another key element is that of primitives. Within some approaches, it was considered that each level has certain primitives. At the conceptual level, primitives are static, discrete symbolic representations. The counterparts of symbolic representations at the neural level are neuronal patterns of activation. Fodor and Pylyshyn insist on saying that the correspondence between the primitives on the conceptual level and those of the neural level is not univocal: „(t)he structure of 'higher levels' of a system are rarely isomorphic, or even similar, to the structures of 'lower levels' of a system" (Fodor and Pylyshyn 1988, p. 63). These patterns of activation are permanently moving and changing.

In this chapter we saw how Fodor and Pylyshyn try to explain these levels. However, their notion of correspondence is different from the one used within the EDWs perspective, being related only to 'levels of description' or to 'layers of reality'.

Other people who worked in various domains have thought about the relationship between levels in different ways. For instance, Clark underlines the co-evolution of representations with processing dynamics (Clark 1997a). Usually, a representation is stored information which stands for something and whose function carries specific information. (Bechtel 1998; Clark 1997b) The systems that use representations accomplish a special kind of coordination with features from the environment: they correlate signals received from the environment with certain inner states that guide behavior. However, the existence of this type of correlations is not enough to establish the representational status of the inner state; the nature and complexity of that correlation are more important. (Clark 1997a)

In order to speak about representation as a 'stand-in' for something, we need to consider the process in which that representation is used. If a process uses a certain representation, then the process must be correlated with a representational format which is a system of conventions for expressing the content of that representation. (Bechtel 1998) What does correlating a process with a mental representation mean, from the EDWs perspective? Does the process belong to the mind-EW, or to the brain (body) that belongs to the macro-EW? For instance, all types of memory belong to the mind-EW. It can be useful for the I to access implicit knowledge, for instance in reactivating mental representations. If we insist on seeing a link between a neural pattern of activation and a mental representation, we create the unicorn-world. Mental representations exist only in the mind-EW, they are the I.

Usually, the debates concerning the notion of representation have focused on the format or nature of representation. Fodor and Pylyshyn argue in favor of the propositional format of representation. Representations must have concatenative compositionality in order to instantiate the productivity and the systematicity of thought. (Fodor and Pylyshyn 1988) Strongly criticizing Quine's holism, Fodor has an atomistic point of view regarding the meaning of a symbol. (Fodor 1998) However, generally speaking, philosophers think that the meaning of a symbol is not intrinsic to that symbol, but it depends on how that symbol is used by an agent or a system. (Heil 2004, p. 114)

What are the mechanisms which produce mental representations and their properties within the mind-EW? For instance, language acquisition requires different psychological mechanisms, such as short and long-term memory, or native mechanisms of learning acquisition that are essential during the critical period of development. However, it is never these mechanisms alone (mental representations and some cognitive processes) that are involved, but also human subjectivity, i.e. the I. We have to return to the idea that the I is both implicit/unconscious and explicit/conscious knowledge.

There are some dynamicists in cognitive science (Globus 1992, 1995; Kelso 1995) who reject not only representations, but also the process of brain computations. Globus replaces the process of computation with constraints that take place between elements and levels of the system, and Kelso mentions that „(r)ather than compute, our brain dwells (at least for short times) in metastable states". (Kelso 1995, p. 62) I consider that both authors are right, but only from the point of view which includes the brain/body (that belongs to the macro-EW perspective). Computation is a notion which describes processes from the mind-EW, not from the brain/body (that belongs to the macro-EW), where the only things that can be directly described are certain processes that correspond to mental states and processes.

When the interactions between the human body and the environment are very complex, certain bodily representations of the external world take over the tasks usually associated with pure innate computational resources, which leads to the increase of behavioral fluency and flexibility. In such situations, the participating processes and states cannot be completely specified, and the behavior does not follow from a set of instructions generated by a general control, but from the self-organization processes that underlie the brain, the body and the environment. (Wheeler and Clark 1999) Clark takes into consideration 'representation-hungry problems' (decision making, counterfactual reasoning, etc.), which involve a potential decoupling between the representational system and the environment, decoupling which is a kind of off-line cognition, rather than the on-line one that dynamicists suggest. He considers that in such cases the cognitive system has to create a certain kind of item, pattern or inner process that stands for a certain state of affairs – in short, a representation. (Clark 1997a)

If the dynamicists strive to describe only the brain/body (that belongs to the macro-EW), they are correct. Indeed, within the brain/body, we do not have any representations or computations, only neural and bodily processes and entities. But this does not mean that there are no representations and computations within the mind-EW. Only the conflation of two EDWs (imposed by the framework of the unicorn-world) and the domination of dynamical processes (which is usually the case for brain-body-world relationships) determine us to completely reject notions such as mental representations and computations within the brain/body.

Dietrich and Markman (2003) oppose four anti-representationalist approaches in cognitive science by stating that cognition must use discrete representations. The main idea is that in order for a system to be able to discriminate inputs, it needs to use discrete representations. (p. 101) In fact, for a system to categorize, it needs to discriminate, and in order to discriminate it requires discrete representations.

Dietrich and Markman offer seven arguments to support their theory: cognitive systems must discriminate between states in the represented world, and they must access specific properties of representations; they must combine representations and have a compositional structure; there must be functional role relations between concepts; and cognitive systems must be able to create abstractions and non-nomic representations.

Dietrich and Markman argue for the difference between continuous and discrete representations. In their critique of other paradigms (such as the dynamical system approach), computationalism (which uses discrete representations to explain cognition) 'is the best paradigm for cognitive science'. (p. 114) However, they lack the EDWs paradigm, which would allow them to avoid the critiques against their approach elaborated by, for instance, the proponents of the dynamical system approach.

As I have mentioned above, both scientists and philosophers have worked within the unicorn-world, which they believe to have different levels. Merzenich and deCharms claim that there is a representational perceptual constancy even though, at the neural level, the pattern of activity of a group of neurons – from which perceptual representations emerge – is permanently changing and moving. As in the case of other approaches, this one is also incomplete due to EDWs. In addition, Merzenich and deCharms take the relations between neurons to be more important than the neurons themselves. This offers a better explanation of the construction of 'novel complex representational combinations' (Merzenich and deCharms 1996, p. 66), which are not directly experienced. This representational perceptual constancy belongs to the mind-EW, and the idea of being 'isomorphic across changing patterns of activity' means nothing more than that a representation corresponds to the changing patterns of activity.

Extending this idea, we can say that the same thing also happens in the case of conceptual representation. Even though the group of neurons that corresponds to a conceptual representation is continuously changing, within the mind-EW we have a conceptual constancy. Karmiloff-Smith proposed the concept of representational redescription, which is similar to perceptual or conceptual constancy. (Karmiloff-Smith 1994) The primitives from the mind-EW correspond to certain primitives and processes from the brain/body (that belongs to the macro-EW). The constitutive entities from the mind-EW, i.e. mental states, play an essential role (we mentioned above Fodor and Pylyshyn's position concerning this idea). Their importance is comparable to that of Bohr's primitive concepts, concepts referring to the macro level: „every image of the world has to be compatible with their existence". (Bohr 1948 in Prigojine 1978, pp. 11–12) These primitives/representations are the I.

We can draw a parallel between two pairs/dichotomies, mind-brain and macro-micro. From Merzenich and deCharms' perspective, at the conceptual level (which is the mind-EW), we have a perceptual constancy, but the pattern of activation that corresponds to the neural level (the brain/body that belongs to the macro-EW) is in continuous motion. We perceive a table as static and discrete in relation to the surrounding environment.[4] Within the micro-EW, a table does not exist; only the corresponding network of microparticles, of which some are in motion, do. The I, as the result of the evolution of the species, has clear, distinct, and complete internal mental representations. Some internal representations (that

4 The shape of the table depends on the environment. In a short period of time and in a standard environment (lacking in dramatic and abrupt changes of temperatures, for instance) the changes are not important.

are the I) correspond to certain external objects such as tables and chairs (that belong to the macro-EW). The changes within the brain/body (that belongs to the macro-EW) are continuous, but there is a threshold of change. This threshold is in fact an organizational threshold of neural patterns of activation. When this threshold is crossed, the neural patterns of activation change their structures. The corresponding entities (mental representations) in the mind-EW are changed. In order to avoid the epiphenomenalist position, I claim that the properties of the I correspond to certain neural patterns of activation. Mental states are not determined by neural states, but only correspond to them. Both EDWs (the I and the macro-EW) have the same objective reality. However, we have to be aware that there are no interactions between the two EDWs, since one EW does not exist for any EDW.

Another important key element is the task. Each scientific theory assumes an explanation of one EW and the construction of a theoretical model. Because of the being/existence of EDWs, when we try to explain various phenomena from a scientific perspective, we have to reject the unicorn-world and to avoid the hypothesis that each element in the universe depends on all other elements. When researchers propose a theory to explain some phenomena, they have to select certain tasks: the explanation of a phenomenon belonging to an EW, which has a particular task. For instance, in connectionism, each network has to solve a particular task. These tasks are merely functional or structural tasks that are different from qualitative states or human subjectivity.

6
EDWs and Cognitive Neuroscience

My last two books (2012, 2014) are dedicated to the research done in cognitive neuroscience in the last 4-5 years. In this chapter, I intend to present a general image of cognitive neuroscience (a sub-domain of cognitive science), followed by an investigation of several essential topics in this field: localization (differentiation–integration in the brain), the binding problem and multisensory integration, and the 'optimism vs. skepticism' debate in cognitive neuroscience as a science. In the conclusion of this chapter, it will become clear that cognitive neuroscience is not a real science, but rather a pseudoscience.

The state of affairs is somehow paradoxical: there have been incredibly many people working in cognitive neuroscience laboratories in the last 20 years, and an avalanche of data has been made available through a variety of research techniques, such as fMRI, EEG, MEG, and TMG. In addition, information is widely shared online, many workshops and conferences take place in the world, but despite all this, the mind–brain problem and other related problems are still unsolved. Therefore, the following question arises: 'Can we solve these problems in the future?' This question is strongly related to another one: 'What kind of science is cognitive neuroscience?'

Paradoxically, even though the majority of researchers in this area accept some form of reductionism, some questions about the status of the mind still exist, despite the fact that in cognitive science, in the main framework, mental states are correlated with neuronal patterns of activation. Nevertheless, these relationships (correlations) create serious problems in cognitive neuroscience. There are important people in cognitive neuroscience (and numerous philosophers) who reject the identity theory. In the philosophy of cognitive (neuro)science, there have been many debates regarding the relationship (or difference) between mental and neuronal states. Is this difference ontological, epistemological, linguistic, or are these states attributes of an unknown substance (as claimed by Spinoza and other people in the philosophy of mind today)?

The identity theory, the idea that the brain produces the mind (Searle 1992) and different kinds of emergence (see Vacariu 2008) are the main paradigms which dominate cognitive neuroscience today. Throughout the last 15 years, there has been a significant progress in cognitive neuroscience. Particularly, the advances in functional neuroimaging technology (fMRI) have offered a real hope that we might understand the relationship between mental states and the activity of neuronal patterns. Nevertheless, even if such investigation tools are very helpful, it seems unlikely that in the near future researchers will finally find the real solutions to the main problems of cognitive neuroscience. On the contrary, the new investigation tools generate more and more controversial answers to the main problems of cognitive neuroscience. Will these empirical results lead to the correct conclusion regarding the main problems of this field of knowledge?

6.1. A general view of cognitive neuroscience

Cognitive neuroscience is nothing more than an accumulation of experimental and theoretical information from two fields, neuroscience and psychology. Various people have constructed different theories based on combinations of these types of information.

The idea of accumulation has to be associated with correlations between neuronal and mental states, which form the main notion of cognitive neuroscience. It is not very clear what the precise meaning of correlations is. On the one hand, many people working in this area argue for the identity theory: a mental state is identical with certain activated neuronal patterns. Nevertheless, many researchers avoid the identity theory. From an epistemological point of view, a mental state cannot be reduced to neuronal patterns. On the other hand, many important researchers in cognitive neuroscience and in the philosophy of mind think that the neuronal states produce the mental states. (The philosopher Searle was the first who introduced this idea, in 1992.) In this way, the problematic identification between a mental state and a neuronal state is partially avoided, since we can speak of a correlation between these states. Nevertheless, correlations require the localization of mental functions/states in the brain. As we will see in this book, because of the results obtained in the past few years (mainly through fMRI), localization is replaced with widely distributed neuronal areas, and correlation remains the main notion of cognitive neuroscience.

I would like to draw attention to the wrong framework in which research is done in cognitive (neuro)science. At a first glance, the old framework of the unicorn world does not have a direct influence on a very large number of empirical experiments and investigations. Nevertheless, their conclusions are either wrong or ambiguous from the point of view of my perspective.

Changing this old framework of thought with a new one could affect not only the answer to some old questions, but even the way in which those questions are phrased. The construction of experimental technologies follows particular frameworks of thinking, so the understanding of measurements/observations achieved with any instrument or the creation of any experiment has to be included within a paradigm of thinking. Based on the knowledge that I acquired by reading articles and books in various fields of cognitive neuroscience, I hope that I will be able to construct, in the next book, the ontological status of the I, the relationship between the I and its features (that are the I), and the correspondences between the mind-EW (the I) and the brain-body (belonging to the macro-EW).

In my future work, I intend to construct a completely new image of the I, which is now more necessary than ever in cognitive neuroscience. This chapter and the preceding one are written for young researchers (who are middle-aged) in cognitive (neuro)science, who have a solid understanding of the information available in this field, and who want to decode the mysteries of neuronal functions and their relationships with particular mental states. Through my investigation, I want to demonstrate that the current directions of research in cognitive neuroscience have no future. For this purpose, I have selected for analysis some of the main topics of cognitive neuroscience.

My bibliography is extremely small, compared to the extremely large number of papers and books published on each topic in this area in the past few years. Nevertheless, in the summer of 2012, I realized that whenever I analyzed a paper from my perspective, the results were very similar: every author works within the unicorn world and the consequences for all of them are the same, i.e., after many experiments and after using extremely elaborated approaches, there is no answer to any of the great questions in this field.

The term 'cognitive neuroscience' was coined by Michael Gazzaniga, who is considered 'the godfather' of this discipline, while talking to George Miller in the backseat of a taxi in New York in the late 70s.[1] (D'Esposito 2010, p. 204) He launched the Journal of Cognitive Neuroscience in 1989, and created the Cognitive Neuroscience Society in 1993. (idem) Almost everybody is aware of the limitations of various technical tools in cognitive neuroscience. D'Esposito (2010, p. 210) illustrated the spatial and temporal resolution of some methods used in cognitive neuroscience in his book.

I want to demonstrate that cognitive neuroscience is just a new kind of engineering, or even worse, a pseudo-science. For this purpose, I will discuss some general notions of cognitive neuroscience used in a textbook written by Baars and Gage and published quite recently, in 2010. I will also refer to details from another textbook (Banich and Compton 2011). These books may be considered standard, up to date books for students who are interested in cognitive neuroscience, since they reflect the current state of cognitive neuroscience perfectly.

1 The label 'cognitive neuroscience' „was coined in the late 1970s in the backseat of a New York taxi when Mike Gazzaniga was riding with the eminent cognitive psychologist George Miller to a meeting to gather scientists to join forces to study how the brain enables the mind.“ (D'Esposito 2010, p. 204)

I want to emphasize some elements which appear within these texts. The main idea which unites them is that the authors underline that, often, the results they present in their books remain controversial.[2] Three elements can be seen as the causes of this state of affairs: firstly, the main subject of research is a very complex entity (the mind/brain); secondly, cognitive neuroscience has only just begun its research; thirdly, the framework used by people who work in this science is wrong. However, regarding the second point, this science has already benefited from 40 years of research and a huge number of researchers, and therefore we can consider that it has a much longer history (not in time, but in empirical results) than other particular sciences. The first and the third points are strongly interrelated: something can be considered complex because of our framework of thinking. Therefore, it is better to consider that the third point is the main cause for the lack of real progress in the theories/approaches in this domain. Working within this wrong framework, I believe that there is no chance for someone to elaborate a general approach accepted by the majority of researchers in the future.

One of the main notions that I analyze in this chapter is localization: the main job of researchers working in this domain is to localize certain mental functions/states in the brain. Even if, in the past few years, more and more people consider that all mental states have a wide distribution in the brain, many researchers have worked to localize each mental state on a large number of neuronal areas of the brain. Against the idea of localization, we have to take into account the notion of 'two-way connections' (that involve re-entrant processes), which appear between nearly any two neural areas.[3] (Baars and Gage 2010, section 3.2) Moreover, it often happens that different neuronal areas work together (for instance, sensory and motor areas, Baars and Gage, section 3.3):

2 I will give some examples from Banich and Compton's book: the area V4 „has been posited to play a special role in color perception, although that claim has been controversial" (p. 161); „despite the continuing controversy in this area" (p. 163); „the organization of cells across inferotemporal cortex is still not fully understood" (p. 182); „researchers are still attempting to understand exactly what inputs the cells use to engage in this computation and how they transform the information from one reference frame to another" (p. 214). „The functional significance of these feedback connections from the cortex is still under investigation…" (p. 153); „Controversy still exists about whether changes in striate cortex activation actually cause the shift in conscious perception between the rival stimuli…" (p. 159); „Does each area—V1, V2, and V3—serve a different function? Do they represent different properties of the visual world? The answer is that we do not really know the functions of all these visual maps." (p. 161). (There are many paragraphs of this sort in Banich and Compton's 2010 book.)

3 „Many cortical regions are massively interconnected with each other, so that activity in one part of the cortex quickly spreads to other regions. A number of scientists believe, therefore, that the entire cortex, together with satellite regions like the thalamus, should be considered as a functional unit. This is often called the thalamo-cortical system." (Baars and Gage 2010, p. 81)

> while there are very clear anatomical divisions between sensory and motor pathways, they are constantly interacting. When we speak, we also hear ourselves speak. When we walk, an array of visual information streams across our retina. Video games that simulate the flow of optical vistas really give a sense of motion, even though they are only visual. The brain is constantly updating its motor systems by means of sensory input, and telling the sensory systems what to expect by way of motor signals. (Baars and Gage, p. 75)

From the EDWs perspective, when we try to localize certain mental functions that belong to the I (better said, they are the I), we must consider that we cannot make a very clear difference between the sensory and the motor pathways. There are a very large number of interactions between these pathways in the brain, and the very clear anatomical divisions between sensory and motor pathways are at most a methodological separation, or in Kantian terms, a transcendental division. At the same time, the mental states which correspond to these pathways are the I, and the simulation of motion through visual games is a consequence of the existence of the I as an EW. Interactions can happen only between the brain-body and the environment, but not between the I and the environment, because the external environment does not exist for the self/I. The 'sense of motion' is the I that merely corresponds, very approximately, to certain neural patterns.

Baars and Gage rely on Fuster's 'classical diagram', which suggests that „the nervous system is always cycling information between input and output channels, to keep the sensory and motor world in synchrony", (Fuster, 2004; Kandel et al., 2004) Fuster describes the hierarchies of local hierarchies inside the brain, which are constantly interacting with the brain, the body and the environment. (Baars and Gage 2010, p. 76) Moreover, these authors mention Friston's hierarchical maps diagram with the feed-forward and the feed-back channels.[4] From an EDWs perspective, when we talk about the hierarchies of hierarchies inside the brain, localization is automatically excluded. We must not forget the fact that localization involves correlations between certain neuronal areas of the brain and particular mental states. What kind of mental states could we correlate with such neural hierarchies of hierarchies? Again, we have to be aware of the fact that localization is very approximate both on the spatial and the temporal scales, since it is an attempt to correlate phenomena which belong to EDWs. The temporal and spatial limitations impose some conditions regarding the correlations of these hierarchies of hierarchies with particular mental states.

Today, fMRI is the main instrument used in attempts to localize mental states. However, I must strongly emphasize that it will be impossible to localize any mental state in the future. To quote one of Baars' and Gage's most important ideas: „(...)

4 „Hierarchies are not rigid, one-way pathways. They allow signals to flow upward, downward, and laterally." (Baars and Gage 2010, p. 91) These bidirectional pathways are against localization and strongly support the EDWs perspective.

neurons do more than fire spikes. The input branches of a neuron, the dendrites, also engage in important activity. By recording different parts of a neuron we get somewhat different measures of its activities." (Baars and Gage 2010, p. 96; my italics) Being aware of this, it is essential that we return to Bohr's principle of including the measurement apparatus in defining the entities and the processes that we investigate: if we use fMRI, then we have to include the structure of MRI technology in defining the phenomena recorded by this instrument. If we use an EEG machine (or any other device for reading/measuring brain activity), then we have to do the same.

Obviously, by using certain conditions of observation, we observe certain entities and processes. When we use internal tools (introspection, consciousness), we observe certain mental states (that is, any mental state is the I.) Using fMRI or PET, we observe the activation of some neural patterns. Using an EEG, we observe certain waves. These states are almost complementary. From my point of view, if we pass an epistemological-ontological threshold by changing certain observational conditions, we observe entities/processes which belong to EDWs. So, it is absolutely necessary to include the conditions of observation in the definitions of entities/processes that we investigate.

FMRI and PET tools have good spatial resolution, but unsatisfactory temporal resolution (for instance, fMRI takes 6 seconds to record a result, due to changes in local blood supply), while EEG and MEG have good temporal resolution. Thus, each apparatus has certain limitations, imposed by its structure, regarding the analysis of the activities of the brain. For instance, it is very important that EEG captures information from the cortex, but not from subcortical regions. These conditions also illustrate that our tools of observation have a priori structural limitations in their recording of phenomena (which also have certain particular structures). In some cases, the results we gain from a set of observational tools are just partial aspects of the phenomena we investigate. Therefore, we must combine the results of various tools of observation/investigation. However, this combination often implies combining contrasting information which belongs to EDWs.

Some researchers combine the results of fMRI with EEG in order to get both spatial and temporal resolutions. Layreys, Boly and Tononi talk about the 'multi-modality integration', which refers to the combination of results offered by two apparatuses, for instance, fMRI and EEG. With such combinations, we get a closer to „complete characterization of the different aspects of the brain activity during cognitive processing". (Layreys, Boly and Tononi 2009, p. 41) In some cases, we have to take into account the principle of complementarity from quantum mechanics, which was elaborated by the famous physicist Niels Bohr. Therefore, it might be impossible to combine certain phenomena (firing neurons and oscillations) characterizing the same entity (the brain in this case). Combining data furnished by different tools could lead us to illicit extensions: the mistake is created when we illegally attribute an aspect (a feature) to an entity. In reality, there are ED features which belong to ED entities.

Another important problem for localization is represented not only by the subcortical zones, but also by the role of neurotransmitters and neuromodulators (which are „classes of neurotransmitters which influence synaptic transmission broadly within neural circuits“, Noudoost and Moore 2011, p. 585) for the corresponding cognitive functions. The roles of these elements are essential for cognitive functions, but the fMRI and the EEG cannot directly record the actions of neurotransmitters/neuromodulators.

> Notice that all of these neuromodulators project axons from small nuclei below the cortex, spreading their neurochemicals to wide regions, both cortical and subcortical. Only a few thousand cell bodies in these nuclei therefore have massive effects in the rest of the brain, controlling sleep and waking, pleasure and pain, alertness and working memory. (Baars and Gage 2010, p. 541)

From an EDWs perspective, the activity of neurotransmitters and neuromodulators (as well as many other types of activity in the brain and the body) must be added to the role of neurons (their properties of firing and oscillations), and it is all of these which correspond to particular mental representations and processes. We cannot ignore any of these elements in explaining the relationships between neuronal states and mental states. However, the problem is not solved yet. The researchers in cognitive neuroscience have made an incredible mistake: they have correlated mental states with particular cortical areas, ignoring the role of subcortical regions. In fact, they have completely ignored the idea that, without these elements, cortical areas would have no contribution (correspondence) to mental processes. That is, while trying to explain the process of any mental state through a correspondence to certain cortical areas, we cannot ignore that any mental state is the I, and it corresponds to the entire brain-body. This is why we have to correlate not only the cortical areas, but also the subcortical regions, the glial cells, the neuromodulators and neurotransmitters to the cortical areas. Amazingly, nobody considers that neuromodulators are as important to mental states as the firing and oscillations of neurons. Why? I believe that this is due to the fact that we found it much easier to construct the tools which record neural firing and oscillations than to construct tools recording the actions of neuromodulators. I really do not understand what the criterion for this separation was.

Baars and Gage illustrate their idea through the experiment made by Quiroga and his colleagues at Caltech (Quiroga et al. 2005 in Baars and Gage 2010) Using the axonal firing rate, they show that a single neuron in the medial temporal cortex is activated only by pictures of the actress Jennifer Aniston. Other neurons were activated by pictures of Halle Berry or the Sydney Opera House. Baars and Gage emphasize the fact that a single neuron can know nothing about Jennifer Aniston.

Usually, there is a large and distributed network of neurons that are activated by a particular face. „If this neuron were lost, the person would still be able to recognize Jennifer Aniston. The brain as a whole would not show a detectable change." (Baars and Gage 2010, p. 96) It is the framework of the unicorn-world which creates this paradoxical problem: the role of that neuron in producing the recognition of Jennifer Aniston. From the EDWs perspective, there is only correspondence (not identity) between mental states and the activity of large neuronal parts which can be recorded using fMRI and PET, to which we must add the activity of glia cells, neuromodulators, neurotransmitters, etc..

Using a new apparatus for scanning the brain, based on the diffusion tensor imaging technique, we can visualize the fiber tracts of white matter which reflect the connectivity across the two hemispheres and four lobes of the brain. Observing the results of a diffusion tractography of white (myelinated) fiber tracts and knowing that the role of the white matter cannot be ignored in explaining cognition, localization becomes very problematic if we only use fMRI and PET. Moreover, even if we accept that particular mental functions are localized in the right and/or in the left hemispheres, the corpus callosum (which consists of 100 millions fibers) integrates information from both sides. (Baars and Gage 2010, p. 134)

This allows us to ask a meaningless question: does the integration take place in the corpus callosum? On the one hand, the two hemispheres are correlated with particular functions (through differentiation or segmentation), whereas, on the other hand, we have an integration of information (for instance, consciousness and even subjectivity have a unity which requires this type of integration). From my point of view, I believe that this pair of notions, differentiation and integration (whole-parts) is a creation of the human mind extrapolated from the macro-EW or micro-EW to the mind-EW in relationship to the brain (which belongs to the macro-EW). This type of distinction is meaningless if we analyze the mind (the mind-EW), since this differentiation requires space or distance, neither of which exists in the mind. (For spatial cognition, see Vacariu 2014, Chapter 2) Moreover, even in the brain, such segmentations are created artificially by human beings who use particular tools of observation (fMRI, EEG, etc.), due to the numerous connections between existing neurons and neuron patterns. Even the notion of 'neuronal patterns' is a very approximate notion which does not reflect something that really exists in the brain. I strongly emphasize the fact that this differentiation (segmentation) is an artificial notion we created, which can only be methodologically and pragmatically (not ontologically) useful, while integration is the I or the mind-EW (with its unity).

I will analyze, from the EDWs perspective, a particular case of integration of spatial information coming from multiple senses (touch, hearing and vision), which was presented in Banich and Compton (2011). This type of integration is called 'multisensory integration'. It seems that visual information is represented by eye-centered representations, while tactile and audible information is represented by head-centered coordinates. (Banich and Compton 2011, p. 214) It is possible that the integration of both kinds of information is done by the parietal cortex. The authors write that the „regions of the parietal lobe manage to align maps constructed from different senses“. (idem) Banich and Compton base their affirmations on a text which contains a concrete example of this type of integration (the superior parietal area aligns both visual and touch information in head-centered coordinates). However, they do not explain what they mean by the phrase 'to align maps', which therefore remains vague. It has a meaning only within the EDWs perspective, in which case it means that it corresponds to a particular mental state.

The central sulcus is a landmark in cognitive neuroscience. It separates the posterior part of the brain from the anterior one.

> Posterior cortex contains the projection regions of the major sense organs – vision, hearing, touch, smell, taste. In contrast, frontal cortex is involved in action control, planning, some working memory functions, language production, and the like. In a sense, the posterior half deals with the perceptual present, while the anterior half tries to predict and control the future. (Baars and Gage 2010, p. 145)[5]

The increasing number of Brodmann areas in the last few decades reflects the framework of the unicorn world. If, at the beginning of the last century, there were around 52 known Brodmann areas, today there are more than 100 such areas acknowledged in cognitive neuroscience. Brodmann areas are correlated with different specialized functions of the cortex (visual, auditory, motor, language, cognition). Nevertheless, as we saw above, the correlation of mental functions with neural areas requires not only activated neuronal areas, but also the activity of white matter and subcortical regions (glial cells, neurotransmitters, neuromodulators and other neuronal processes).

> While the cortex is vital for cognitive functions, it interacts constantly with major satellite organs, notably the thalamus, basal ganglia, cerebellum, hippocampus, and limbic regions, among others. The closest connections are between the cortex and thalamus, which is often called the thalamo-cortical system for that reason. (Baars and Gage 2010, p. 127, italics in the original)[6]

5 „The basal ganglia have been implicated in action planning and unconscious cognitive operations. New evidence, however, has linked the basal ganglia to higher order cognitive functions, such as decoding the grammar, or syntax, of language.“ (Baars and Gage 2010, p. 151) „Unlike regions in sensory cortex, the frontal lobes do not have a single job to do – they are not specialized for decoding speech sounds or recognizing faces. Rather, the frontal lobes are engaged in almost all aspects of human cognitive function.“ (idem, p. 400) These sentences directly support the EDWs perspective!

6 „Think of the thalamus as a relay station: almost all input stops off at the thalamus on the way to cortex; almost all output also stops off at the thalamus, going out to the muscles and glands. Fibers emanating from cortical cells spread in every direction, flowing horizontally to neighboring cells, hanging in great bundles on their way to distant regions of cortex, and converging downward on the

Due to its two-way or re-entrant connections, the thalamocortical system is deeply interlinked and active. (Baars and Compton 2010, Chapter 8, section 1.5) From an EDWs perspective, the role of the thalamo-cortical system becomes essential in understanding the notion of correlations. Can we still find correlations within the thalamo-cortical system? Each mental state is associated with a neuronal area which is more activated than the others, as well as with other neuronal areas working in a more silent manner (see Vacariu 2005, 2008), but to these we have to add neuromodulators, neurotransmitters, white matter and subcortical areas, in other words, we need to add the entire brain (and body, see Sporns 2006 or Vacariu 2008). This is an application of the 6 Principle (the being corresponds to an organism/cell). The quote above refers to a huge approximation (by naming the 'thalamo-cortical system'), but in reality it is about the entire brain (and the body). Analyzing Edelman, Tononi, and Sporns' works, Baars and Gage conclude that

> Recursive complexity reflects the fact that brains show rich organization at multiple levels, from molecular interactions within individual synapses to reentrant interactions among distinct brain regions. A combination of measures may be needed to adequately characterize the neural dynamics relevant to consciousness. (Baars and Gage 2010, p. 292)[7]

From the EDWs perspective, this type of complexity and levels do not exist in nature, because nature itself does not exist. Complexity is the most complex Ptolemaic epicycle in biology and cognitive science, created so as to reflect the pseudo-relationship between the mind and the brain. Any mental state (which is in a

great traffic hub, the thalamus, of each half of the cortex. In addition, hundreds of millions of axons flow crosswise, from one hemisphere to the other, creating white axon bridges called commissures." (Baars and Gage 2010, p. 149) And we also have to keep in mind that „the thalamus is the major input hub for the cortex, and also the major cortex-to-cortex traffic hub, like a large airport that might serve both domestic and international traffic. However, the basal ganglia operate as a major output hub, for motor control and executive functions. The brain has multiple hubs, just as it has multiple superhighways." (idem, p. 250; see also Chapter 8, section 1.7) It seems that the unity of the mind automatically involves such re-entrant connections. Taking these re-entrant connections and numerous hubs into account, what do we really observe using fMRI, PET or EEG? Do we have any chance to correctly localize any mental function? I will refer to Uttal's words: „As the discussion in this book progresses it will become clear that modularization and localization are no longer tenable interpretations." (Uttal 2011, p. 43) Moreover, we have to take into account the changes regarding neurons and their interactions during the development and evolution of species!

7 „Some theorists suggest that the entire cortex, or the thalamo-cortical system, should be viewed as massive networks for integrating, differentiating, and distributing signals (e.g. Edelman, 1989; Edelman and Tononi, 2000; Freeman, 2004)." (Baars and Gage 2010, p. 394) You will recall my analogy between a table and microparticles, and the unity of mind/consciousness and neural patterns of activation. I will refer to Uttal's words: „There are too many uncertainties, too many neurons, too many idiosyncratic interconnections (e.g., the brain is not neatly organized as is a simple crystalline structure) for us to ever be able to understand its detailed organization and how, specifically, this complex information pattern produces the reality we call mind." (Uttal 2011, p. 29)

temporal framework, the I) has to be correlated not only with the posterior and the frontal brain parts, but also with the entire cortex and the subcortical areas (which are within a spatio-temporal framework). However, we must keep in mind that the internal time is also the I.

In the spirit of Kant's transcendentalism and Bohr's principle (we have to include our tools of measuring in the definitions of entities/processes), through our use of various apparatuses, we institute a spatio-temporal framework which will frame the neural patterns of activation. The question which we are then faced with is: which are the criteria by which we can establish a correct correlation between a spatio-temporal neural state/function and a mental temporal state/process? For instance, the „time lag between the two hemispheres working on the same task may be as short as 10ms, or one hundredth of a second (Handy et al., 2003)" (Baars and Gage 2010, p. 140) What does 10ms or 50ms mean for the I? The minimum time necessary to visually perceive an external object is 10ms. Is this limit imposed by the time lag? We have to take into account that there are ED times for the brain-EW (which is the brain that belongs to the macro-EW) and the mind-EW. The I is not the sum of 10ms brain states, but an indeterminate entity that has a unity which cannot be found within the spatio-temporal frameworks imposed by our apparatuses of observation.

Again, applying Bohr's principle of including the structure of an apparatus used to observe a phenomenon in the definition of that phenomenon (see Vacariu 2008), we also have to include the structure of an fMRI or EEG in defining the entities/processes observed with these tools. It is believed that 100ms represents 'the minimum conscious integration time in perception': two discrete sensory processes are integrated into a single conscious event if they occur in an interval of maximum 100ms. (Baars and Gage 2010, p. 289) After several experiments, Doesburg and his colleagues concluded that lateralized visual signals lead to gamma phase-locking, lasting around 100ms, across large areas of cortex. (idem) Due to this, it becomes impossible to correlate a particular mental state/function with certain neural areas.

Another element which argues against localization is the spontaneous activity of the brain, or the intrinsic brain processes. (The neuroscientist Marcus Raichle was among the first who proposed this notion in 2001, see Baars and Gage 2010, section 3.2; for more details about the 'default network', see Vacariu 2014) If we take this intrinsic activity of the brain into account, then the most activated neural patterns which we record using fMRI are only the top of the iceberg, but the entire iceberg corresponds to the I. Each mental state/process is the I. We are again confronted with the integration-differentiation problem, or with the whole-parts relationship, which often appears in Baars and Gage's book. For instance, the basic visual features are color, orientation, motion, texture, and stereoscopic depth. (Baars and Gage 2010, p. 159) Thus,

> … most neurons in early visual areas of the brain are highly tuned to specific features – some may fire very strongly to a line shown at a particular angle, to a particular color, or to a particular motion direction. These neurons respond to a very small region of the visual field (i.e. your current field of view), ranging from just a fraction of a degree to a few degrees of visual angle. … If the activity of each of these neurons represents only a small part of the visual field, such as whether a small patch of the visual field contains vertical or horizontal, red or blue, motion or something stationary, then how is the brain able to combine this information across many neurons? Somehow, the brain is able to organize these basic feature elements into organized perceptual groups. (Baars and Gage 2010, p. 159)

On page 394, they claim that

> Chapters 3 and 6 showed that the visual system appears to have at least one region of integration 'where everything comes together', the inferotemporal cortex (area IT) (Sheinberg and Logothetis, 1997). In that area, neurons respond not to single retinal stimuli, nor to separate features like colors or light edges, but rather to entire visual objects. It is at least possible that language may have a similar region of integration (Hagoort, 2005; Figure 11.28). (Baars and Gage 2010, p. 94)

It seems that we have another hub, which is a point of integration. Undoubtedly, these hubs are similar to the (in)famous homunculus from the philosophy of mind. Baars and Gage bring into discussion the Gestalt principle. In their words, the „German word, Gestalt, is difficult to translate directly, but expresses the idea that the whole is greater than the sum of the parts". (idem, p. 159, italics in the original)

This paragraph is the best reflection of the wrong framework, the unicorn world, in which the researchers elaborate their empirical experiments and approaches. In reality, we are dealing with EDWs and nothing else. The neurons which respond to a particular stimulus are the 'tip of the iceberg', while the 'organized perceptual groups' are the I, so in this case we are confronted with at least two EDWs.[8]

I will now analyze the relationship between the I, perception, the brain and the external world from the EDWs perspective. (See Vacariu 2011) As I have showed in other works, the external world has to be incorporated within the I, in the Kantian sense.[9] There are several points here that need to be stressed. As I showed in Vacariu (2011), perceptions do not exist, since they belong to the mind-EW (more precisely, all perceptions are the I). Only the brain interacts with the

8 Moreover, we have to take into account that different cognitive functions can access the same neural area (one-to-many) and that the same mental function can instantiate at least partially different neural areas. (Uttal 2011, p. 55-57)

9 As opposed to this, as we saw in the previous chapter, cognitive science relies on different approaches (such as the dynamical system approach, or Clark's extended mind) which promote the extension of the mind into the external environment. (See Vacariu 2008, and Vacariu and Vacariu 2010)

body and the external world, but, it is meaningless to talk about interactions between the mind-EW and the external world (the macro-EW) because, in such cases, we are actually faced with two EDWs.

In this context, I would like to analyze two notions. The first notion refers to the 'constructive perception' or the 'perceptual filling in view'. I consider that perceptual filling is one of the most important topics in cognitive neuroscience: our eyes have a blind spot because they lack photoreceptors at the back of the retina, where the axons of the retinal ganglion are unified to create the optic nerve. (For 'constructive perception', see Baars and Gage 2010, section 5.4) However, we do not notice this because the brain „fills in perception of the blind spot". (Baars and Gage 2010, p. 186) Moreover, such fillings occur not only for the blind spot, but also for other parts of the visual field. (idem) In those areas, the brain fills in color, patterns and motion. It is only at the center of our gaze that we perceive in full color and high resolution. (Baars and Gage 2010, p. 158) The fovea „subtends about four degrees of visual arc" (idem, p. 48).

> Outside of the fovea, which covers only 2 to 4 degrees of arc, the retina loses resolution and is sensitive only to light and dark edges. (It follows that our normal sense of a rich and colorful visual world is a construction of the visual brain, not a literal record of the input into the retina.) (Baars and Gage 2010, p. 272)

Obviously, this is an extremely important observation. How can we draw the line between what we really perceive and what is filled in by the brain? How do we explain the way in which the information that is really perceived by the eye and the brain combines with the information that is 'filled in by the brain' (or by the mind)? From an EDWs perspective, these questions are meaningless: all the perceptual/conceptual/psychological information belongs to (or, better, it is) the mind-EW. It is neither a combination between the external and the internal worlds (of information), nor an extension of the mind outside the body (as embodied cognition, etc.). In the Kantian sense, the external world (more exactly, the 'perceptual world') is incorporated into the mind (not into the brain). The brain (including eye-movement) interacts with the external environment, but the mind does not.

The second topic I wish to discuss refers to the fact that in the occipital lobe, in the V1 area, images have 2 spatial dimensions (they are 2D), while the I perceives 3D images. It is then logical to consider that we are not aware of what happens in V1, even if V1 is sensitive to visual features like orientation, direction of motion, colors and color differences. We have 3D visual perceptions, but where does the brain construct them? V1 projects feedforward signals toward V2, V3, V4 (which are important for the perception of color), and toward the MT area (which is important for motion). (Baars and Gage 2010, p. 166) Moreover, it is widely accepted that these projections are divided in two pathways: the ventral pathway (from V1 to the temporal lobe – which is important for representing what objects are) and the dorsal pathway (from V1 to the parietal lobe – which is important for representing where the objects are located).[10] (Fig. 7)

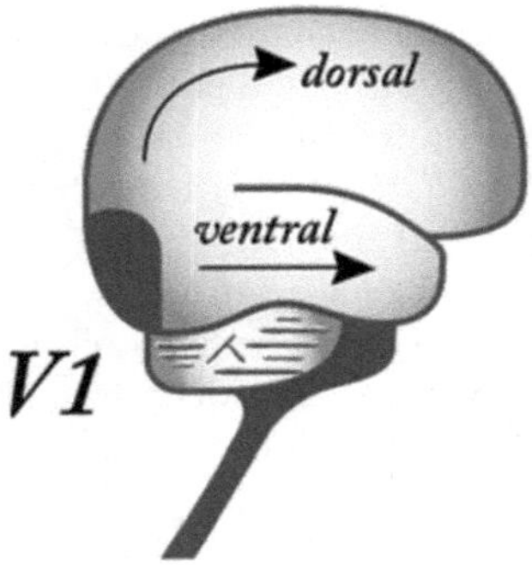

Fig. 7 Two main streams.
The ventral stream: from the primary cortex (V1) to the temporal cortex; and the dorsal stream: from the primary cortex (V1) to the parietal cortex. Obviously, these streams are just idealization: any part of the brain is directly connected to many other parts.

Nevertheless, the authors mention that the dorsal-ventral distinction is not an absolute one, since there are many 'cross talks' between them, and the parietal and the temporal lobes send projections to the prefrontal cortex area „where information from each pathway can also be reunited". (Baars and Gage 2010, p. 169) From my point of view, the dorsal and the ventral pathways are very rough approximations which very largely correspond to features of mental representations. The what and the where features of a perceived object are in the mind-EW and cannot be found anywhere in the brain.[11] Uttal wrote that it is important „to

10 Here, Baars and Gage mention Ungerleider and Mishkin (1982), Goodale and Milner (1992) and others. (Baars and Gage 2010) Ungerleider and Mishkin (1982) showed that „two distinct streams were postulated: a temporal or ventral stream devoted to object recognition and a parietal or dorsal stream devoted to action or spatial tasks. Ungerleider and Mishkin termed these the what and where pathways." The first one is the „ventral stream, V1 → V2 → V3 → V4 → IT (…)" and the second is the „dorsal stream, V1 → V2 → MT → MST." They go on to explain that „[t]he dorsal stream is dominated by magnocellular cells and the ventral stream by parvocellular cells, although the segregation is far from strict (Merigan and Maunsell, 1994)." (Reid and Usrey 2008, p. 657)

11 Related to where and what, Weiskrantz's famous blindsight is another example that supports the EDWs perspective. His patient, who was visually impaired due to cortical damage, could still point to

remember that no matter how complex the analysis, brain images essentially search only for answers to the where question. The essence of the mind-brain problem, however, is still the how question and it is not yet clear just what the where question tells us about the mind-brain problem". (Uttal 2011, p. 46) From the EDWs perspective, 3D visual perceptions are the I, or they belong to the mind-EW, and therefore it is meaningless to search for the area where these constructions take place in the brain. The entire brain and body correspond to the construction of 3D visual perceptions and all other perceptions that are the mind-EW.

Baars and Gage indicate that, according to Ganis et al (2004), almost the same neural patterns are activated for both visual imagery and visual perception (these patterns are in areas from the occipital, temporal, and parietal cortex). (Baars and Gage 2010, p. 48) This is another argument in favor of EDWs: visual perception is an EDW rather than the macro-EW where the brain/body is situated.[12] Baars and Gage suggest that the right posterior parietal cortex is a possible neural area, from which the 'perspective of the self' emerges. (p. 293) Nevertheless, they write that there are „still many unanswered questions about endogenous brain rhythms and how they interact with external inputs". (Baars and Gage 2010, p. 298)

Another great problem in cognitive neuroscience is memory.[13] I believe that it is impossible to localize the parts of the brain that are responsible for a particular type of memory, or for memory in general. Baars and Gage consider that memory storage implies „very widely spread synaptic alterations in many parts of the cortex", even if the hippocampus is responsible for the transformation of the experience in both memory and map for spatial localization. (p. 306) The medial temporal lobe is another hub with widely spread connections to many areas, like the visual, auditory, somato-sensory, emotional, motor, memory, and executive areas.[14] Therefore, this lobe receives, binds, and distributes information for the long-term memory. Obviously, there are great debates in cognitive neuroscience about localizing conscious events, working memory, and selective attention. (Baars and

the location of the light source. Baars and Gage consider that „[t]hese findings suggest that there can be dissociations between visual processing in the brain and a person's subjective awareness". (Baars and Gage 2010, p. 177) Obviously, from an EDWs perspective, this happens because of the presence of two EDWs.

12 Baars and Gage point out that exactly as visual imagery corresponds to visual perception, there is an inner speech that corresponds to outer speech. (Baars and Gage 2010, p. 49) Again, this is another argument for the being of the I.

13 Attention is another important problem for cognitive neuroscience. Since we are talking about EDWs (mental and neuronal), almost all correlations produce important and still unsolved problems in cognitive neuroscience.

14 „All brain regions are involved in learning, memory, and plasticity, which can be considered as different methods for evoking long-lasting adaptive changes in the brain." (Baars and Gage 2010, p. 541) „Learning occurs everywhere in the brain. Memory storage occurs in the same regions that are used in active tasks." (idem, p. 542) Lost in localization!

Gage 2010, p. 309) Baars and Gage are of the opinion that working memory might overlap with attention, conscious events, and episodic recall. (p. 337) From my point of view, we have to be aware of the fact that all these processes are the I, and therefore their precise localization is quite impossible. Paradoxically, the progress of research in cognitive neuroscience (mainly through the creation of various mechanisms of investigation, such as fMRI) do not solve any problem. On the contrary, the already existing problems become more complicated and new problems appear.

6.2. Optimism in cognitive neuroscience

One of the most important philosophers in cognitive (neuro)science, Bechtel, was an optimist, even if a couple of years ago he admitted that the process of perceiving a simple object had to be correlated with more than 30 neuronal areas (Bechtel 2008). Bechtel's approach is the 'mental mechanisms': „A mechanism is a structure performing a function in virtue of its component parts, component operations, and their organization. The orchestrated functioning of the mechanism is responsible for one or more phenomena." (Bechtel & Abrahamsen, 2005; Bechtel, 2006) (Bechtel 2009, p. 6; 2008, p. 13) Working within the unicorn world, this definition breaks the parts-whole rule. This is why we have to re-think Bechtel's mechanisms within the EDWs perspective. Bechtel was convinced that the localization (and decomposition) of mental states in the brain would be successful in the future, pleading for the heuristic theory of identity (Bechtel 2008).

Later, Bechtel considered that the notions of localization and brain areas needed to be re-conceptualized (Bechtel 2013). His new alternative is a combination of mechanisms with the dynamical system approach, i.e., dynamical mechanisms, and the explanation of the „endogenous activity of the brain" (Bechtel 2013). He believes that in order to explain cognition through neuronal processes, we need to clarify this intrinsic activity of the brain (related to Raichle's default network). Bechtel already tried to combine reductionism with emergence (Bechtel 2008) and integration with differentiation of operations (Bechtel 2009). In order to support these ideas, Bechtel refers to Sporns and Zwi's (2004) 'dual role of cortical connectivity': the functional specificity of certain cortical areas that manipulate specific information, and the integration of this kind of information (Bechtel 2013). Bechtel insists on combining integration with parallel localization of certain various functions. Bechtel wants to preserve not only decomposition, but also the autonomy of a system by introducing Bernard's notion of 'internal environment' (Bernard's expression in Bechtel 2009, p. 12 or Bechtel 2008) or Cannon's homeostasis and its extended notion, Varela's autopoiesis. (See Vacariu and Vacariu 2010)

> Autonomous systems are mechanistic (dynamic) systems defined as a unity by their organization. We shall say that autonomous systems are organizationally closed. That is, their organization is characterized by processes such that (1) the processes relate as a network, so that they recursively depend on each other in the generation and realization of the processes themselves, and (2) they constitute the system as a unity recognizable in the space (domain) in which the processes exist (p. 55). (Bechtel 2008, p. 217)

In 2009, Bechtel adds that

> In fact, living systems are typically highly integrated despite the differentiation of operations between different organs and cell types. The mind/brain seems to be no different on this score – it consists of component processing areas that perform different computations which are nonetheless highly integrated with each other. Such a mechanism does not typically include encapsulated modules, and one is not likely to find them in the mind/brain. (Bechtel 2009, p. 15)

From my point of view, I notice that functional specificity and integration are quite similar with the binding problem and localization. I investigated Bechtel's mechanisms in many of my works, but mainly in Vacariu and Vacariu (2010), Vacariu (2012) and Vacariu (2014). From my point of view, Bechtel's biological mechanisms really exist, but they do so either in EDWs or in the same EW (for instance, the macro-EW where the brain and body are, but, in this case, the mechanism does not exist for its parts, nor vice-versa). If the entire mechanism and its parts existed, there would be an ontological contradiction: it is impossible for two (sets of) entities to exist in the same place at the same time. From my point of view, decomposition, an important notion in Bechtel's mechanistic approach, does not exist, it is a concept created in our mind. Moreover, if the mechanism does not exist for its parts, it similarly does not exist for other entities/processes that interact only with the parts. It is clear that Bechtel's mechanisms really exist, but not within the unicorn-world. I have discussed the fact that EDWs can provide the hyperontological background for Bechtel's mechanisms in my 2012 book.

Expanding on the optimism paradigm, we can consider that the work of researchers from Gallant's laboratory produced the best results in cognitive neuroscience in the past few years, as represented in the following papers:

(a) Nishimoto et al. (2011): „Reconstructing Visual Experiences from Brain Activity Evoked by Natural Movies“
(b) Huth et al. (2012): „A Continuous Semantic Space Describes the Representation of Thousands of Object and Action Categories across the Human Brain“
(c) Stansbury et al. (2013): „Natural Scene Statistics Account for the Representation of Scene Categories in Human Visual Cortex“
(d) Çukur et al. (2013a and 2013b): „Attention During Natural Vision Warps Semantic Representation across the Human Brain“ (with the fusiform face area as an example)

I will investigate the first great realization achieved by Gallant's team.[15] In 2011, Nishimoto et al. published an article about a new method for 'mind reading' (this work being considered among the most important achievements in the last 15 years in cognitive neuroscience) (Nishimoto, Vu, Naselaris et al. 2011). With the aid of a computer program which processed fMRI results, the researchers constructed a quantitative model of brain activity. Using brain activity measurements, Nishimoto et al. reconstruct the natural movies seen by three human subjects. It is the first study which reconstructs dynamic stimuli (natural movies) based on brain activity, with the aid of fMRI. In the past, only static pictures were reconstructed, the main problem being that the blood oxygen level-dependent (BOLD) signals measured by fMRI are much slower than the neuronal activity in relationship with dynamic stimuli.

The researchers of Gallantlab focus on signals received by the early visual neural areas V1 (the functionality of this neural area being quite well studied), V2 and V3 (all three areas being part of the occipitotemporal cortex lobes). The measured training data of brain activity (BOLD signals evoked by 7,200 color natural movies, each movie presented once) is used to match an encoding model for each voxel from the posterior and the ventral occipitotemporal visual cortex. Then they use a Bayesian decoder to reconstruct movies from the evoked BOLD signals, i.e., combining the „estimated encoding models with a sampled natural movie prior, in order to produce reconstructions of natural movies from BOLD signals" (Nishimoto, Vu, Naselaris et al. 2011, 1642). Comparing the fMRI data and the details of each movie, the computer program constructs dictionaries for shape, edge and motion. Each voxel has such a dictionary. The subject watches a second set of movies and new fMRI data is collected. Using the computational models constructed after the first set of movies, the second set of movies is reconstructed only from the second fMRI data (Nishimoto, Vu, Naselaris et al. 2011).

Using the 'pick a card, any card' magic trick, the people working in Gallantlab do not even try to offer an alternative to the binding problem.[16] Even if Uttal admires Gallantlab's work, he concludes that all they were able to do was to show distinctive fMRI responses from a number of visual cortical areas (V1, V2, V3, V3A, V3B, V4, as well as the lateral and anterior occipital cortex), which could be used to identify images from the training set. What they did not do was to take fMRI images and use them to directly plot the pictures of the original stimuli; once again, they selected pictures from their library, based on the pattern of activations. This is not reconstruction per se, but a selection from a predetermined 'deck of

15 For all the other research done by Gallant's team, see Vacariu (2014).

16 In their previous works, they write that this problem „is analogous to the classic 'pick a card, any card' magic trick" (Uttal 2011, 113).

cards' (Uttal 2011, 114). Therefore, Gallantlab's works do not offer any real alternative to the mind-brain problem.

6.3. Skepticism in cognitive neuroscience

Uttal, who is not a philosopher, but a researcher in cognitive neuroscience, is emblematic for the contemporary skepticism regarding the localization of certain mental functions through imagistic procedures. His main book against localization was published in 2001, but Uttal furthered his ideas in a book published in 2011. Even if he accepts the identity theory, he provides many arguments against localization.

In an ontological postulate, Uttal considers that mental processes are the results of interactions which happen at the micro-level of the brain. Since fMRI and PET localize mental functions at the 'macro-level' (the large neural patterns), the results are completely wrong (Uttal 2011, 11). Through a corollary of this postulate, Uttal believes that 'the neural network approach is computationally intractable' and thus the mind–body problem cannot be solved (Uttal 2011, 26). Moreover, he subscribes to the general view in cognitive neuroscience which claims that the „brain activity associated with mental activity is broadly distributed on and in the brain" (Uttal 2011, 45).[17]

17 „All brain regions are involved in learning, memory, and plasticity, which can be considered as different methods for evoking long-lasting adaptive changes in the brain" (Baars & Gage 2010, 541). Lost in localization! (There is a paper written by Derrfuss Jan and Mar A. Raymond (2009), „Lost in localization: The need for a universal coordinate database", NeuroImage, vol. 48, 1–7; see Vacariu 2012) Anyway, this widely distributed neuronal network correlated with any mental state needs to be linked to Baars' global workspace (Baars & Gage 2010), which is supported by Dehaene's global neuronal workspace and by Raichle's default mode network: „But, just as we will not understand evoked activity without first understanding intrinsic activity, so we will not understand consciousness without first understanding non-conscious activity, for in both instances the latter dominates the former" (Raichle 2011, 155). Haynes also believes that Baars' global workspace theory seems to be the best alternative, since many experiments have shown that the distributed areas are involved in each mental task (Haynes 2009). Following the same route of holism, Bressler and Menon strongly argue that cognition is much better explained at the level of „large-scale networks" (Bressler & Menon 2010). Baars specifies other important people from cognitive science who adopted the 'global workspace' theory: Edelman (1989), Damasio (1989), Freeman (1991), Llinás et al. (1998), Edelman, Tononi (2000), Kanwisher (2001), Dehaene, Naccache (2001), Rees (2001), John (2001), Varela et al. (2001). (in Baars and Franklin 2007) So, this holistic position is against localization: the activation of large parts of the brain is correlated with any mental state/function. A large number of people adopted the holistic view concerning correlations. (For details about the holistic view in cognitive neuroscience, see Vacariu 2012, Chaper 11) However, there are many researchers who try to localize particular mental functions in a very limited neuronal area (that is, localization).

Here, we can observe an epistemological-ontological framework- which shows us that neural networks are indeed 'computationally intractable'. We can find no mental computations within the brain. Uttal believes that localization through fMRI and PET is the wrong method of identifying mental states. He claims in his first epistemological postulate for neuroscience that the „brain activity associated with mental activity is broadly distributed on and in the brain. The idea of phrenological localization must be rejected and replaced with a theory of broadly distributed neural systems accounting for our mental activity" (Uttal 2011, 45). For Uttal, the main reason for this is that the actual tools operate at the wrong 'level of analysis' and the mind would be better understood not at the macroscopic level, but the microscopic one. What is important is that „a priori no macroscopic brain imaging or electrical recording activity, no matter how direct it may seem to be in recording the activity of the brain, can in principle provide solutions to the mind–brain problem" (Uttal 2011, 26). Moreover, „many different cognitive processes can activate the same area or system of areas of the brain" (Uttal 2011, 55) and „many different regions of the brain are activated during any kind of cognitive task" (Uttal 2011, 45).

As opposed to localization, Uttal is almost entirely certain that any mental task/function/state (sensation, perception, simple thought) involves the entire brain. He claims that all the parts of the brain are somehow interconnected; it is not possible to isolate the neural patterns which correspond to any cognitive process; consciousness and all related processes (thinking, reasoning, decision-making, problem solving, and intelligence) are the most problematic notions in cognitive neuroscience; etc. There are no clear definitions of some mental states, like emotion, attention or consciousness[18], and such states are probably general functions, not modules of cognition. Conclusively, we can suggest that Uttal's perspective is the strongest type of skepticism in cognitive neuroscience. Actually, Uttal considers that he promotes realism in cognitive neuroscience, not skepticism. (personal correspondence)

Another already classical example of questioning fMRI can be found in the works of Vul et al. (Vul, Harris, Winkielman et al. 2009). Vul et al. investigate the correlations between the behavioral and self-report measures regarding personality or emotions, and the measurements of brain activation obtained using fMRI. They show that these correlations are higher than should be expected given the (evidently limited) reliability of both fMRI and personality measures. (Vul, Harris, Winkielman et al. 2009, 274) The inquiries of Vul et al. concerning the questions and methods of fMRI are reported in 54 articles. The authors of these papers try to find empirical data in order to bridge the divide between mind and brain: extremely high

18 For instance, Uttal quotes Vimal (2009) who offers „a list of 40 different meanings of consciousness and argued that even this list was not exhaustive" (Uttal 2011, 271).

correlations between measurements of individual differences relating to personality, emotion and social cognition, and measurements of brain activity obtained using fMRI. (Vul, Harris, Winkielman et al. 2009, 274)

I will not go into detail regarding this investigation, but I will reproduce its conclusion: Vul et al. claim that such correlations are 'impossibly high', in other words, extremely exaggerated and far from reality. Even if Vul et al. urge the authors of the articles which were investigated to correct the results of the correlations made based on very few empirical results, I believe that these correlations will never be perfect.[19] Today, the most important and the most widely used method of investigating the brain in order to explain the mind is neuroimaging (mainly non-invasive fMRI, but also PET, MEG, etc.). The neuroimaging tools help the researchers in cognitive neuroscience to associate particular neuronal areas with cognitive functions.

One of the problems with such associations/correlations is that we cannot be sure that the cognitive process we associate with certain neuronal areas is totally isolated from other cognitive processes (D'Esposito 2010, 207). As a result, observed neural activity may be the result of some confounding neural computation that is not itself necessary for the execution of the cognitive process seemingly under study. (D'Esposito 2010, 208) As D'Esposito emphasizes, no method in cognitive neuroscience is perfect.[20]

6.4. Localization and the binding problem

Two of the main problems in cognitive neuroscience today are localization and the binding problem. When a subject is in a mental state, fMRI is used to scan his brain and to attempt to localize the neural patterns responsible for that particular state. A notion that has been considered to be related to localization (especially in the last 3-4 years), is that of integration: the identification of some neuronal processes which are correlated with the unity of consciousness, or the binding processes. There have been a large number of experiments in cognitive neuroscience under the umbrella of localization (segmentation or differentiation)-integration (binding). Nevertheless, localization remains the main topic of cognitive neuroscience.

19 Uttal mentions a large number of people who draw attention to the (theoretical and empirical) limits of fMRI (Uttal 2011). In a very recent article against blobology (Poldrack's expression for the localization of functions in blobs), Poldrack pleads for an absolute methodological rigor in using fMRI (Poldrack 2011).

20 Another important problem in cognitive neuroscience is Raichle's 'default network' (Raichle 2011), or the „dark energy of the brain" (Raichle 2006, Raichle and Mintun 2006), which I investigated in another part of this book.

In this subchapter, I will analyze the main ideas concerning segmentation/localization in visual perception (one of the most important topics in cognitive neuroscience). Various researchers attempt to convince us of the power of localization provided by the neuroimaging instruments (fMRI, PET, etc.). Nevertheless, I want to show that, from an EDWs perspective, these tools can offer us only very approximate information about the correspondences between particular neural areas and mental states.

I will refer to the very interesting experiment conducted by Seymour et al. (2009), which refers to the localizations of color, motion and conjunction between color and motion. In the beginning of their paper, they mention that both features are processed by distinct (even if connected) neural areas: color is connected to blobs in V1, thin stripes in V2 and V4, while motion is connected to layer 4B of V1, thick stripes in V2, and V5/MT.2 Thus, motion and color seem to be segregated at the cellular level, lesion studies confirming this segregation: lesions in V4 impair color perception, but spare motion perception, lesions in V5/MT impair motion perception, but spare color perception. (Seymour et al. 2009, p. 177)

If there is such a functional segregation, how and where does the binding of the above-mentioned features take place?[21] Seymour et al. conducted an interesting visual experiment in which human subjects perceived two transparent stimuli in motion, each stimulus being composed of two rotating circles of the same two colors. The stimuli differed in the direction of the rotation: in one case, one colored circle moved clockwise, the other counterclockwise; in the other case each color was paired with the other's motion. During the entire time, their brains were scanned with fMRI. The authors specify that „the double-conjunction stimuli would be indistinguishable without conjunction-specific responses, as all four feature-specific units are active in both conditions". (Seymour et al. 2009, p. 178) To quote Whitney on Seymour et al.'s paper:

> There were two double conjunction stimuli, both of which contained the same feature information (red, green, clockwise, and counterclockwise). The only difference between the two double conjunction stimuli were the pairings of color and motion: in one, red was paired with clockwise motion and green was paired with counterclockwise motion; in the other, red was paired with counterclockwise motion and green was paired with clockwise motion. (Whitney 2009, p. R251)

The main conclusion of their experiment shows that the primary visual cortex includes information not only about the direction of the motion and the color hue,

21 Seymour et al. wrote that it „remains a matter of debate whether visual-feature binding is mediated by a temporal code (…), by communication between visual areas (…), by feedback connections to early visual areas (…), or by representations at higher, cognitive stages (…)." (Seymour et al. 2009, p. 117) Eternal questions!

but also about the conjunction of these two features. (Seymour et al. 2009, p. 180) „Whereas some areas showed better performance as well as biases for decoding one feature over the other (e.g., V5/MT+ for motion; V4 for color), information about both features and their conjunction was present in nearly every visual cortical region." (Seymour et al. 2009, p. 180) More precisely, the authors of this work also emphasize the limits of the spatial-temporal resolution of fMRI. Moreover, Whitney emphasizes some problems with Seymour et al.'s 'clever technique': beside lesions, we have to take into account the psychological and physiological models of binding based on higher-level mechanisms; also, it is possible that the feedback from the fronto-parietal attentional region helps create the conjunctions; under the illusions' framework, it is possible for the mechanism of feature binding not to be recruited for unambiguous visual stimuli. (Whitney 2009, pp. R252-R253)

I will now analyze Seymour et al's conclusions from an EDWs perspective. My intention is to bring further criticism to this work. As I wrote above, Seymour et al. are aware of the limits of fMRI and PET. If we add to them the role of neurotransmitters, the feedback from other neural areas, and the theories which say that many parts of the brain are involved for any mental task (Baars's global workspace, Edelman's re-entrant processes, Raichle's default network, Libet's CMF, Uttal's view, or synchronized oscillations), can we really hope to localize particular visual features with precision within V1, V2, V3, V4 or V5? How can we integrate localization/segmentation and integration in such conditions?

Again, we have to recall Bohr's method of introducing the conditions of observation into the definitions of neural and mental entities/processes. So, it is very clear that by using fMRI and PET, we can record the neural features of the brain specific for each neuroimaging tool. The question is what the criteria to attribute color, motion and the conjunctions of these features to V1 or V2 are. Could we correlate certain perceptual features only with the firing of some neurons? But the 'neurons do more than fire spikes'. (Baars and Gage 2010, p. 96) From the EDWs perspective, we have to be aware of the fact that the brain and the mind (which includes perception) are EDWs and therefore we cannot assume that only V1 or V2 are correlated with color, motion and their conjunction. Introspection, consciousness and the I are also involved in such experiments, so we would also have to find the neuronal activities that correspond to these entities.

Through neuroimaging, we only get very rough approximations of localizations. These localized neural patterns correspond, with large approximations, to particular mental states, but these mental states are not observed by the I, they are the I (not parts of the I). The occipital lobe is indeed dedicated to vision, but this lobe also corresponds to the I and thus the localization of the exact visual mechanism becomes very problematic. Indeed, according to Derrfuss and Marr (2009), „we are lost in localization".

One of the most important issues in cognitive neuroscience is the binding problem.[22] Some authors claim that we need to solve this problem in order to understand the functioning of the mind and the relationship between the mind and the brain.[23] There are various definitions of binding, but the classical one (perceptual binding) refers to the relationship between certain activated neural patterns in different parts of the brain, which are correlated with the various features/properties of an object (color, size, motion, orientation), and the unity of mental representation in the case of that object. This definition concerns the distinction between segmentation and integration (which is partly related to consciousness).

Different authors propose various types of binding: spatial and temporal binding, part binding (mentally segregating the parts of an object from the background and binding them together), conscious and unconscious binding, perceptual (the unification of perceived aspects) visual binding (linking together color, form, motion, size, and position for a perceptual object, or binding various perceptual objects), auditory binding, sensory-motor binding, crossmodal identification, and memory reconstruction. All these types of binding can take place within a single modality, or across modalities, for example in the case of sensory-motor integration, in cross-modal binding, in action control, or across perception and action, etc. I will continue by analyzing the perceptual visual binding process, the best-known kind of binding.

The binding problem clearly reflects the relationship between the parts (different areas of the brain) and the whole (the unity of mind or consciousness, or experiential subjectivity). Obviously, it is strongly related to the framework created by the mind-brain problem and the problem of representation. Inevitably, in order to solve the binding problem, we need a framework for the mind-brain problem. However, even if we accept the identity theory, we cannot solve the binding problem: we cannot identify a mental state with one or more neural patterns of activation. This is the hardest problem for people working in cognitive neuroscience simply because it is a pseudo-problem.

The main task of cognitive science is to solve the binding problem by using modern technology (fMRI, PET, MEG, etc.). An essential idea in this line of research is that binding happens „almost everywhere in the brain and in all

22 In fact, all the very important problems of cognitive neuroscience (for instance, multisensory integration, a particular case of the binding problem, see Vacariu 2014) are very strongly related simply because many of them are pseudo-problems constructed within the unicorn world!

23 Roskies (1999) considers the binding problem to be „one of the most puzzling and fascinating issues that the brain and cognitive sciences have ever faced". Triesch and von der Malsburg (1996) regard the binding problem as one of today's key questions about brain function. Treisman (1996) indicates that a „solution to the binding problem may also throw light on the problem of the nature of conscious awareness" (Velik 2010, p. 994).

processing levels". (Velik, p. 994) Nevertheless, if we stopped believing that the identity theory is correct, and if we changed the framework, the solution to the binding problem would arise immediately. In my framework, the binding problem is merely a pseudo-problem because this binding does not exist in the brain. There are only certain neuronal processes and entities/patterns in the brain, but not this integration. The integration is actually the I, which is an EDW from the brain/body.

The binding problem is strongly related to mental representation[24], another very problematic notion in cognitive science and in the philosophy of mind. Without mental representations, our species would not survive in its evolution. From my point of view, we have to accept that mental representations really exist, otherwise psychology would lose its status as a science. There have been strong disputes concerning the status of representations, which inevitably involves the status of mental functions. (See Vacariu et al 2001 or Vacariu 2008) The process of binding refers to certain mental attributes which represent the properties of, for instance, a visual object. The arguments for the existence of binding processes are, for example, synesthesia (its most commonly known variation is grapheme-color, i.e. adding color to black letters), or Balint's syndrome („difficulty localizing different objects or object parts, changing their direction of gaze, or shifting their focus of attention from one aspect of the scene to another" and simultanagnosia – the inability to perceive more than one object at a time). (Schmidt 2009) There are some people who consider the binding problem a pseudo-problem, or a problem which can be avoided.

I will investigate today's most accepted variation of the binding problem: the synchrony or temporal binding „formulated independently by C. Legendy in 1970, P. Milner in 1974, and C. von der Malsburg in 1981" (von der Malsburg 1999, Velik 2010).[25] The binding processes are realized through synchronous neuronal oscillations at different frequencies. Using EEG or MEG, the electrophysiological signals are obtained at the level of the scalp, and they mirror the 'synchronization of weak synaptic currents' which are „summed across a large number of neurons; scalp signals therefore necessarily reflect synchronized neural activity" (Tallon-Baudry 2009, p. 322) or the 'rhythmic modulation of discharge activity (neuronal oscillations)'.

The coupling of neurons through synchronization depends on adjusting the phase relationship or the frequency of cells from that neuronal group. The phase of an oscillation furnishes the window for processing information. The inputs in the 'good phase' of the ongoing oscillation are selected, whereas the inputs in the 'bad phase' are suppressed. (Moser et al. 2010, p. 199) One hypothesis is that the syn-

24 For more about mental representation, see Chapter 5 of this book or Vacariu (2008).

25 I will not investigate an old alternative, Triesman's feature-integration theory (FIT) (see Vacariu 2012), or more problematic alternatives, such as binding by convergence or by population coding.

chronization between different areas is achieved through zero-phase lag between the same frequency-oscillatory activities. When two brain regions fire in similar ways, but with a time lag, the term phase locking is more accurate than synchrony. Sound waves echoing in a canyon are phase locked, but not synchronous, because they echo back and forth with a brief time lag. Because neurons also need time for their axonal spikes, related brain regions echo each other with a time lag, leading to phase locking rather than synchrony. Both synchrony and phase locking are commonly observed in the brain. (Baars and Gage 2010, p. 252)

The results of several experiments have shown that the zero-phase lag synchronization can occur in local brain areas or over large distances (between hemispheres, Engel, König, Kreiter et al. 1991), even if there are great conduction delays, due to the pathways which connect the synchronized neural groups. (Moser et al. 2010, p. 205) We must recall that EEG results record only the surface of the waves; underneath what is visible to the EEG, there are various kinds of interactions among waves (locked in synchrony with each other, phase-locked, transiently coordinated, cross-frequency coupling) with different ranges, (recent discoveries indicate ranges from 0.01 to 1000 Hz). (Baars and Gage 2010, p. 254, Chapter 8) Nevertheless, Baars and Gage emphasize that „brain rhythms are a moving target, as new evidence appears with remarkable rapidity". (Baars and Gage 2010, p. 261)

The main topic of a very recent book edited by Philips, von der Malsburg and Singer (2010) is the dynamic coordination within the brain/mind. I would like to draw attention to the abstract of von der Malsburg's paper from 2010:

> Trying to apply our everyday concept of coordination to the brain raises a number of fundamental questions: What is the nature and meaning of local brain states that are to be brought together? On what grounds are they to be coactivated and connected? What is the nature of meaningful structural relationships, and how does the brain learn them? What is the role of focal attention? How does the brain assess its current level of coordination? How do brain states address goals? What is the nature of our environment's statistics, and how is it captured by the brain? What mechanisms endow brain dynamics with a tendency to fall into coordinated states? Some of these questions seem to be difficult to address within the current experimental paradigm. (von de Malsburg 2010, p. 149)

This abstract directly reflects the situation of cognitive neuroscience today. In fact, we already saw other authors asking the same or very similar questions. Thus, it seems that the progress of local knowledge (mainly in neuroscience) does not offer any general direction of research (or working framework) in cognitive neuroscience. Actually, this progress does not exist.

Referring to the work of several authors, Tallon-Baudry shows that there are disparate results regarding the correlations of the occipital, temporal, parietal and frontal regions, and regarding the focal activation confined to the occipital pole. The problem is that MEG and EEG offer us different images of the visually induced oscillations: „while EEG data reveal a short-lived burst of oscillatory synchrony between 30 and 60 Hz and 200 and 300ms, MEG studies consistently report sustained oscillations at higher frequencies". (Tallon-Baudry 2009, pp. 322-3) The results show that visual stimuli produce gamma oscillations in different areas and at different frequencies. (p. 324) Tallon-Baudry strongly emphasizes that there is no strict correspondence between a frequency band and a cognitive process. (p. 239 or 2009, p. 325) There is no simple one-to-one relationship between a frequency range and any cognitive function. (Tallon-Baudry, 2009) After at least two decades of extensive research on synchronized oscillations, nobody can claim that a frequency band is responsible for (correlated to) a particular cognitive function. Moreover, the functional role of oscillatory synchrony in distinct frequency bands may simply depend on the functional specialization of the area that generates these oscillations (Tallon-Baudry et al. 2005), much as the functional significance of ERPs depends on the areas that generate them. (Tallon-Baudry 2010, p. 240) Since it is quite difficult to observe synchrony oscillations in detail, their role in binding processes is unclear and still controversial. What is important to notice is that this hypothesis is about how binding is signaled, not about how binding is computed. (Velik 2010, p. 997) Moreover, synchronization cannot be an alternative to the enduring trait of representation of an object (LaRock 2010, p. 455 or 457). Indeed, 'we are lost in the binding problem.'

6.5. Cognitive neuroscience: science, or 'new engineering'?

In each 'special science' (Fodor 1974) (for instance, physics, neuroscience, psychology), we can find different theories/approaches (quantum mechanics, Einstein's theory of relativity, Fodor's LOT, various neuronal perspectives) which deal with particular entities (micro- and macro-particles, neurons and mental representations) and laws.[26] Such particular entities have questionable (relative or

26 In a famous article from 1972, Anderson (who won the Nobel Prize for physics) shows that reductionism does not explain some physical phenomena appropriately (Anderson 1972). That is, the explanations/theories of some macro-physical phenomena cannot be reduced to quantum mechanics. In the philosophy of mind, we can take into account Fodor's article (published two years after Anderson's paper). If, in Anderson's case, we can talk about a kind of organizational non-reductionism, Fodor somehow establishes a linguistic non-reductionism. Each special science (for instance, neuroscience or psychology) has its own taxonomy that cannot be reduced to basic science (physics) and we cannot mix the taxonomy of neuroscience with that of psychology (Fodor 1974). Special sciences exist not because „of the nature of our epistemic relation to the world, but because of the way the world is put together: not all natural kinds (not all the classes of things and events about which there are important, counterfactual supporting generalizations to make) are, or correspond to, physical natural kinds" (Fodor 1974, 439). Much more recently, Piccinini wrote that „when it comes to explaining cognitive capacities, computational explanation is proprietary to psychology—it does not belong in neuroscience" (Piccinini 2006, 343).

not) ontological status. Cognitive neuroscience deals with correlations which reflect the relationship between entities (mental and neuronal states) with questionable ontological status. Leaving aside the necessity to provide arguments proving this, it is quite clear that correlations have no ontological background, not even a questionable/relative ontological background. Thus, cognitive neuroscience has no ontological entities and no laws. Therefore, in a standard framework (in which we define all the other particular sciences, such as physics, neuroscience, cognitive psychology), cognitive neuroscience is not a real science, but a pseudo-science created through a mixture of information which describes entities/processes belonging to different 'special sciences'. In this context, it is clear that the enormous amount of correlations of the last decade almost led to an informational congestion in the case of young researchers.

I believe that it is possible to regard cognitive neuroscience as a kind of 'new engineering'. However, we can make an analogy between brain imaging (the main tool of cognitive neuroscience today) and neural networks or connectionism (a popular approach between 1990 and 2005). Even if, since the end of the 1980s, connectionism has been very important in cognitive science, researchers' interest in neuronal networks strongly declined in the last few years (for more details regarding connectionism, see Vacariu 2008). I believe that brain imaging is in a similar situation: there is a lot of enthusiasm for it today, but its popularity will decline tomorrow. In other words, I predict that brain imaging or mind reading will have the same trajectory as connectionism: mind reading (in particular) and cognitive neuroscience (in general) will become merely a kind of new engineering. The main reason for this prediction is that brain imaging does not really explain cognition, nor the true relationship between mind and brain.

In my previous two books (2012, 2014) I analyzed several other very important topics in today's cognitive neuroscience: localization, the binding problem, multisensory integration, spatial cognition[27], perception and object recognition, etc.

27 In Vacariu (2014, Chapter 2), I investigated 'spatial cognition' in relationship with the EDWs perspective. I analyzed some of the main sub-topics related to spatial cognition: retinotopic maps, spatial navigation and cognitive maps, hippocampus, grid cells, head direction cells, border cells, egocentric and allocentric representations, frames of reference, integration, the endurance problem, abstract space, 'perceptual filling' and the 'panoramic view', parallel space, sensory modal interactions, color, language, visual mental imagery and visual perception. Different authors try to explain spatial cognition through „abstract spatial representations" (Chafee and Crowe 2013, Wolbers et al. 2011, etc.), but I will refer to the ideas expressed by two persons who worked extensively on spatial cognition: „yet the complex interplay of spatial information processing, perception, action, and higher-level reasoning involved in their use is still poorly understood". (Gunzelman and Lyon 2011, p. 743) It is clear that the thought framework for people working on spatial cognition is the unicorn world, and I will make an analogy between color and space in order to explain this situation. Do we have the green color in our brain or in our mind? We only have a mental representation of an object that it is associated – due to the frequency of light – with the green color. The same thing happens in

None of these problems has a viable solution. From my point of view, all problems in cognitive neuroscience are pseudo-problems (i.e., neither of these problems has a solution). Therefore, I believe that the status of cognitive neuroscience is that of a 'No ontology landscape'. Bassett and Gazzaniga (2011) end their article with this conclusion:

> Neuroscience desperately needs a stronger theoretical framework to solve the problems that it has taken on for itself. Complexity science has been posited as a potentially powerful explanation for a broad range of emergent phenomena in human neuroscience (…). However, it is still unclear whether or not a program could be articulated that would develop new tools for understanding the nervous system by considering its inherent complexities. (p. 208)

This paragraph refers to cognitive neuroscience. If one of the fathers of this domain concluded an article in this way, it is clear that cognitive neuroscience has a questionable status. After the above observation, the article continues with some questions regarding the research directions in cognitive neuroscience today. As an alternative to this problematic situation, the authors introduce the complex theory. From my EDWs perspective, the complex system theory would create, as it did in biology (see Kauffman's theory, 1995, 2000, 2008, as well as our interpretation in Vacariu and Vacariu 2010), a landscape of analysis with many hyperontological contradictions.

Gazzaniga introduces the same analysis in a quite interesting paper published in 2010. The protagonist of this very short paper (written in a creative, fictional style) is an extraterrestrial who, after landing on Earth, analyzes the work of cognitive neuroscientists in their laboratories (they are trying to explain the mind and the brain through neuronal information). Gazzaniga refers to the relationship between matter and life, mentioning Mill[28] and the irreducibility of emergent properties (indicating Broad's work in 1925). The ET remarks that, unlike these two authors, neuroscientists 'dislike this type of thinking', believing that the

the case of space and macro-objects: space, just like the green color, exists neither in our brain, nor in our mind. Notions such as the abstract representation of space or supramodal properties can only be properly understood within the EDWs perspective. Exactly as we have an abstract representation of the green color, we have an „abstract representation of space". The same thing happens for the object recognition process: „The ability to recognize objects across so many varying conditions implies that at some level, our mental representation of objects is fairly abstract, in the sense of being removed from or independent of the original stimulus conditions." (Banich and Compton 2011, p. 190) We have to extrapolate this idea not only from the green color to space, but also to every perceptual property of any 'external scene', that is, we have to follow Kant and to situate the external world (i.e., all perceptual representations of the 'external world') inside the self. (see Waxman 1995) Inevitably, this movement pushes us to change our old framework of thinking, the unicorn world, with the new one, the EDWs.

28 „(…) no mere summing up of the separate actions of those elements will ever amount to the action of the living body itself." (Mill „A system of logic" (1872), qtd. in in Gazzaniga 2010, p. 291)

understanding of the neurological parts would offer a full explanation of the psychological states. (Gazzaniga 2010, p. 291) The problem is that the mind emerges from the brain and produces the feeling that a ghost could 'sneak into the brain'. On the contrary, other thinkers from physics, chemistry, biology (Anderson (1972) is mentioned, but see Vacariu 2008) used the notion of emergence without any problems. In this context, Gazzaniga asks the following

> So how does the brain do it? Understanding how each and every neuron functions still tells us absolutely nothing about how the brain manufactures a mental state. Sure, they all conduct electrical impulses and secrete neurotransmitters in the service of communication. But how does this produce thoughts and feelings? And how can this system keep ticking after the interacting neurons are disrupted by structural or metabolic lesions? Just as a social democracy continues to work when component individuals are eliminated, so too does this biologic network. It is as if the emergent function guides the underlying physics. (Gazzaniga 2010, p. 291)

If the godfather of cognitive neuroscience asked such questions more than four decades after its appearance, then something is quite wrong with the working/thinking framework of those who work in this interdisciplinary field. This framework is, obviously, the unicorn world. From my point of view, all these questions are pseudo-problems within the unicorn world. Gazzaniga analyzes more difficult notions from the philosophy of mind and cognitive (neuro)science, such as levels of analysis, bottom-up causality, the bidirectional causality between micro- and macro-entities, and, most of all, emergence. The ET remarks that the neuro-scientists already started accepting emergence. It is worth quoting its last words:

> ET's spaceship was waiting for him. As he boarded the craft, he mused that the report to home base would be easy. „The earthlings are stuck in a quagmire. They don't see that brains are decision-making devices and should be understood in those terms – that level of description, not lower. They are only partially evolved. It will be eons before they ever find us. It might also be eons before they ever understand themselves." (p. 292)

From my point of view, this ET is an entity who landed in the unicorn world, who noticed the hypercontradictions discovered by the living entities of that strange world. The ET tries to convince them that they are coming up with overly complicated explanations for quite simple EDWs, but obviously nobody believes him.

Cooper and Shallice (2010) draw attention to the fact that if cognitive neuroscience focused too much on neuroscientific data and thus disregarded psychological data, there would be two consequences: this domain would no longer bring decisive contributions to cognitive science, and the neurological knowledge would become meaningless. (Cooper and Shallice 2010, p. 403) While mental, neuronal and physical states generally have an ontological status, correlations do not, lacking even a questionable/relative ontological background.

Cognitive neuroscience is a fascinating, captivating area of research, but the reality (the EDWs) does not accept the mixture of knowledge and ontologies which belong to different special sciences, which study EDWs. Like connectionism, cognitive neuroscience is not science, but only a kind of 'new engineering'. Ever since Descartes, people working in the philosophy of mind and in cognitive science have searched for answers in the wrong place (the unicorn world). The unexpected movement (see the first motto of this book) that these people must make is to reject the unicorn-world and to accept the EDWs perspective. Through the blows dealt by Copernicus, Darwin, Freud and other thinkers, the human being has been losing this dictatorial position in the world. However, with the appearance of EDWs, the old position of human beings is entirely lost, since even the world, the Universe, does not exist. The second feeling they will have will be that of freedom, because they will reject many pseudo-problems in cognitive neuroscience and other special sciences.[29]

In the context of the EDWs perspective, we have to be aware that the 'point of view' of a neuron (i.e., its interactions with other neurons and glial cells) creates another EDW than the one we observe using our perception, the fMRI, and any other tool of investigation. Moreover, the parts of the neuron, such as RNA, DNA, molecules and proteins have their EDW other than that of a neuron. Therefore, the number of EDWs increases dramatically. It is not the job of a philosopher to identify the number of EDWs. This is the job of scientists who work in special sciences within the EDWs perspective. The final verdict: 'We are lost in cognitive neuroscience.'

29 „That's what's so nice about empiricist cognitive science: You can drop out for a couple of centuries and not miss a thing." (Fodor 2001) Fodor's statement mirrors the empirical and theoretical research in cognitive neuroscience in particular much better than it does cognitive science in general. „Looking for consciousness in the brain is like looking inside a radio for the announcer." (Nassim Haramein)

7
EDWs and Biology

In this chapter I will analyze the main unsolved problem in biology: life and its relationship with any organism (or cell[1]). Life and mind (the subjectivity, i.e. the I) have the same ontological status. The I (living being) represents not only the mind and subjectivity, but also life.[2] We cannot really differentiate between mind (the I) and life. Clark (2001) highlights the analogy between mind and life taken from Godfrey-Smith (1996a, p. 320) in order to suggest the strong continuity between them:

> Life and mind have a common abstract pattern or set of basic organizational properties. The functional properties characteristic of mind are an enriched version of the functional properties that are fundamental to life in general. Mind is literally life-like. (Godfrey-Smith 1996 in Clark 2001, p. 118)

Extending this likeness, we can say that mind, human subjectivity (the I/mind) and life should indeed be included in a single concept, that of the living being. In the language of the unicorn world, we cannot consider that a cell has cognition, but everybody assumes that life emerges from a cell. According to my terminology, a cell and a human organism correspond to life, which is the living being, therefore the notion of emergence is wrong. There is a perceived difference between the mind, as an EW with specific entities and processes (the mental states/processes which are the I), and life, which is an EW lacking any such states or processes. This perceived difference appears because the old notion of life can correspond either to an organism (the whole of multi-cellular parts), or to a unicellular organism (or any 'cell of a body'), while the old notion of mind corresponds only to a brain (a large number of cells which are a part of the body).

Therefore, the difference is given by the corresponding complexity of the organism (the number of cells and their connections), but this difference is entirely artificial. In reality, these meanings are created by human beings within the unicorn world. It is not compulsory for every EW to have any particular states/processes.

1 Biologists consider that the cell is the smallest entity which possesses life.

2 Obviously, in this section dedicated to life, I will refer more to the living being instead of the mind/the I, but for me, they are equivalent notions.

The feeling of pain is the I (only from the 'first-person perspective'), but there is no 'feeling of life'. The I does not perceive life in any way, because the I is life/mind. The difference is that pain is an entity/process of the mind-EW (even if pain is the living being/the I), while life has no such entities. We can claim that the manifestations of life are different from those of the mind, but life is mind-like.

Precisely like in the example with the car (or the table) and its components, the cells do not exist for an organism, nor vice-versa, since even if both the cells and the organism belong to the macro-EW, they do not do so at the same time. However, we have to avoid breaking the 'parts-whole' contradiction (the whole and the parts cannot exist in the same place at the same time) and therefore the cells do not exist for the organism and vice-versa. An organism, as a whole, has no parts – the whole and the parts cannot exist in the same place at the same time. From the point of view of the whole (an organism), the parts (cells) do not exist. The whole corresponds to those parts. Any part of an organism/cell[3] is just a mental construction (if the organism really is broken into its component parts, it will cease to exist as a whole).[4] Essentially, as we will see below, even if the organism and the cells exist in the same EW, they cannot exist there at the same time.

I can draw a parallel between two pairs of notions: self-body and table-microparticles. From our point of view, a body is composed of cells. We already know that each organism corresponds to the self/life or to its cells. Then, the question that arises refers to the nature of the relationships between the life which corresponds to the organism as a whole and the life which corresponds to each cell. Obviously, no biologist would admit that the life of an organism is composed from the lives of all the cells that compose the organism. However, this answer is constructed within the unicorn world, therefore it is wrong. The question: 'What is the relationship between the life of an organism and the life of each cell of the organism?' is constructed within the unicorn world, so it is meaningless. Since there is no space within the mind or within life, the 'unity-composition' distinction is meaningless when it comes to the unity of the mind-EW and its components (various perceptions and thoughts).

3 I will generally refer to an organism but, in many cases, those statements are also valid for any cell, because biologists consider that the cell is the smallest living entity on Earth. As we will see below, I believe that there are no major differences between the features of the mind, or life, and the I or the self. Therefore, within the EDWs perspective, these notions (the I, self, mind, life) refer to the same thing. Any cell or organism corresponds to life, the mind, or the self.

4 Even if the parts-whole distinction has no ontological status, we have to continue using it in everyday life, due to its utility. However, in some cases, this distinction produces anomalies and contradictions which can be avoided only with the EDWs perspective.

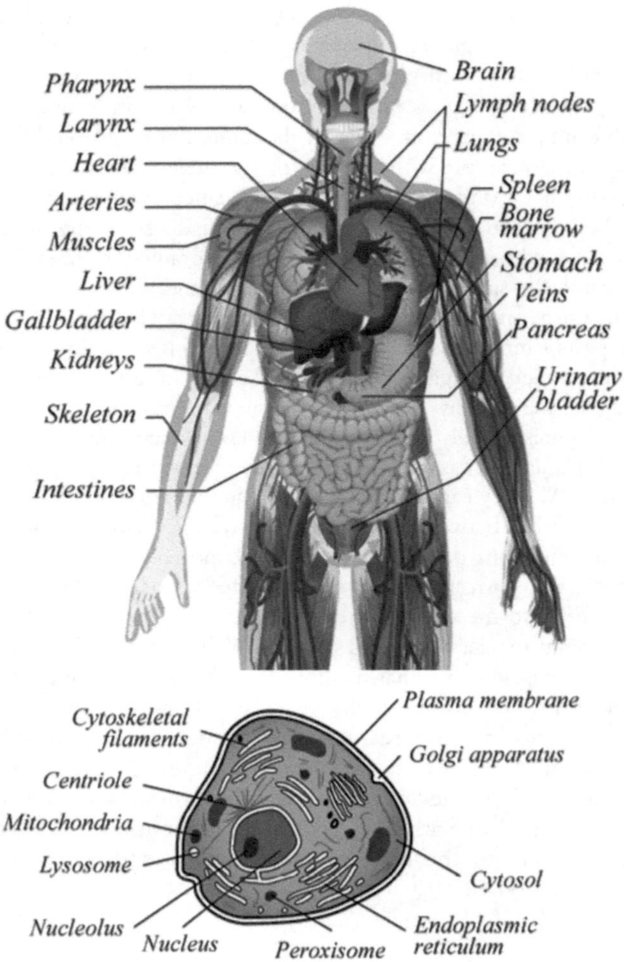

Fig. 8 The cell/ organism and life
Where is life? Above, we can see the depictions of a cell and an organism and their components: (1) The cell/organism-life relationship: the cell/organism belongs to the macro-EW, its life is an EW. Life does not exist for the cell and vice-versa. (2) The whole-parts relationship: the whole does not exist for the parts, i.e., ontologically, the cell/organism cannot exist in the same place, at the same time as its components.

From the EDWs perspective, the state of affairs is as follows: since life is an EW, the life that corresponds to each cell does not exist (or, better said, it cannot be) for the life that corresponds to the organism; the life of each cell does not exist for the lives of the other cells, even if one cell exists for all the other cells. However, the organism does not exist for its cells. (Obviously, we are confronted again with the whole-parts problem: the whole does not exist for its parts and vice-versa). Each different life, the organism, and the cells are or belong to EDWs. (Fig. 8)

The parts-table dichotomy differs from the body-self dichotomy, but its conclusion is ultimately the same. Even if the parts and the table both exist in the same macro-EW, they cannot exist there at the same time. The extension of the table is not caused by the extension of its parts. If we consider that a table is composed of certain parts, then we reach the 'parts-whole' ontological contradiction. The extension of the parts does not exist in the same place and at the same time as the extension of table. Therefore, this decomposition is possible only as an abstract thought (it can exist in our mind, but not in reality).

Within a single EW, the parts-whole relationship leads us to an ontological contradiction. Again, it is meaningless to try to find the relationship between the I/living being which corresponds to the organism and each living being which corresponds to each cell. This is simply because each living being is an EW, therefore one living being does not exist for any other living being. The composition or decomposition of living beings is absurd, since any living being (life/mind) is an indeterminate individuality (it has no spatial dimensions). It would be very strange to state, for example, that 'This living being is composed of many living beings'. In fact, the notion of living being (the self/mind) has no plural. The living being does not interact, so it is again meaningless to look for the relationship between the being which corresponds to the organism and the beings which correspond to the cells. The living being (the I/self/mind) simply corresponds to an organism.

If we stated that 'The living being exists' (not as it is correct, „The living being is"), it would mean that we were searching for such relationships and spatial dimensions. By saying this, I do not refer to the limits of human observation/perception/thought, but to the status of the indeterminate individuality of the living being. Therefore, the living being/the I is (not exists). Above all, we have to keep in mind that we must not break the Kant-Carnap rule and that we must not reach the 'parts-whole' ontological contradiction. The composition of the I/self/life is a notion that has no meaning. Can we talk about the composition of life? The life that corresponds to an organism cannot interact with the life of another organism, since life is an EW.

We have already seen what the relationship is between an electron and a table. The situation is the same when we refer to life and a cell. We can perceive a cell (using, for instance, a standard microscope), but life is a subjective experience/the I that corresponds to the entities/processes of an organism/cell. We would need a 6 sense to perceive life, but, as we have seen earlier in this book, such an instrument or ability is impossible to obtain even from a theoretical point of view. Nonetheless, such a sensorial mechanism would be different from our perception, or from a standard microscope. In other words, we cannot perceive the cell and life with the same instrument, at the same time. Life/the I is an EDW from an organism which belongs to the macro-EW. Like the I, life neither emerges from, nor is produced by the body. Life is nothing more, nor less, than the living being. We can explain an organism/cell which corresponds to life, but there is no causality between life and the organism/cell, only correspondence.

The huge efforts made throughout the history of human thought in order to explain cognition or life were all placed within the framework of the unicorn world (where EDWs were mixed), which led to the creation of many hybrid models. (For details, see Vacariu and Vacariu 2010) Such hybrid models are necessarily heterogeneous, mixing elements which belong to EDWs. Obviously, these hybrid models are possible only within the unicorn world. We have to be aware of the fact that we cannot explain cognition or life by introducing the hybrid models. The EDWs perspective does not cope with hybrid models. Kant avoided this heterogeneity (created by mixing the doctrine of rationalism with the principles of empiricism) by bringing the world (more precisely, the cognitive perceptions of the world) inside the I. The EDWs perspective also avoids the possibility of any such heterogeneity, and this is the main reason why the notion of the world has to be excluded from our language.

The main topic in biology is the life of cells/organisms. Neurons are one of the 265 different types of cells in our body. From the human point of view, the mind is an EW which corresponds to the brain and the body. Since the notion of the mind is included within the notion of life, I can assert that life can correspond to the organism (both the brain and the body), or it can correspond to a single cell. The similarity is that nobody can localize either cognition or life within a spatial framework. An EW cannot be localized in a spatio-temporal framework, only the entities/process that belong to some EDWs can. Not all EDWs have a spatio-temporal framework in which their entities can exist. In the actual context of cognitive science and biology, the difference between mind and life is this: if something is life, it does not necessarily have cognition. On the contrary, cognition needs to be life. For me, life is equivalent to cognition.

In this context, I would like to change our perception of this relationship. From an EDWs perspective, we need to anthropomorphize even the cells: any organism/cell corresponds to intelligence. There is a difference of degree (but not a natural one) between the mind that corresponds to the cell and the mind that corresponds to the human being. Human cognition implies many more functions than the cognition of a cell, or that of a primitive organism. The first cognition corresponds to many more entities/processes than the second one. However, cognition and life have the same ontological status. Having this common status, we could really extend the EDWs perspective from cognition to life. Life is an EW that corresponds to an organism/cell. Accepting that the cell is the primitive entity of life, the life of a cell corresponds to its molecules, membrane, DNA, RNA and proteins. The life of an organism is an EW, and the organism belongs, from a human point of view, to the macro-EW. If the I is knowledge (the implicit/unconscious and the explicit/conscious knowledge), life is a kind of implicit knowledge. Without this type of an implicit knowledge, that is, without innate knowledge (it corresponds to the evolution of life and to that of the species), even a cell would not survive in its environment. Moreover, this implicit knowledge represents the subjective unity (the I) which corresponds to a biological cell.

If one accepts that the Big Bang took place 13.7 billion years ago, one will also agree to the appearance of living entities from non-living matter, and to the evolution of species. In the 19th century, Darwin elaborated his theory of evolution, which became largely accepted in the last century. If, from the human point of view, but within the unicorn-world, we analyze the entities and the processes which appeared after the Big Bang, we have to assume a sort of continuity regarding the evolution of all classes of entities, from microparticles and macroparticles to cells and animals.

In other words, it is a continuity between matter, life and cognition, or between objects, organisms and life/the mind/the I. We have to be aware of the fact that the framework of this continuity is the unicorn-world. Therefore, from the EDWs perspective, we need to eliminate this continuity. Since we eliminated the continuity of appearance between microparticles and macroparticles (there is no relationship between them since one set of objects does not exist for the other set of objects), and the one between the mind and the brain (the mind does not exist for the brain, nor vice-versa), we have to eliminate the continuity between matter (i.e., body/cell) and life as well. Obviously, the life-EW (the mind-EW) does not exist without the corresponding matter. Nevertheless, there is no causality (continuity) between matter and cognition, so there is no continuity between matter/body/cell and life. Cognition/life and matter are, or belong to, EDWs, so there are no relationships (either causality or laws of complexity) between them, but only some correspondences.

Within the unicorn world, it has been quite difficult to find a definition of life. There is a widespread belief that life is somehow related to certain biological/physical processes. In fact, life has been attributed to physical organisms which can be localized within a spatio-temporal framework. Temperature, which corresponds to the movement speed of molecules/atoms, is likewise a process, but at least it is one that we can physically observe with the help of our senses, or of certain artifacts. Life is neither an abstract, nor a perceptual notion. Then, what is life?

Even within the unicorn world, it is a tautology to say that life is a property of a living entity, but this abnormality seems to be quite normal within the unicorn world. Since cognition and life have the same ontological status, we could claim that life does not emerge from cells, just like cognition does not emerge from neurons. Life is an EW which corresponds to an organism/cell which belongs to the macro-EW. An organism corresponds to its cells and their processes, and a cell corresponds to its molecules (DNA, RNA, proteins, enzymes and their chemical processes). It is a contradiction to consider that a cell is identical with, or composed of, or emerges from, its molecules and their activity.

Many researchers (led by Kaufmann) tried to explain life within the framework of complexity. While trying to discover the laws of complexity, Kaufmann insists that one of the main principles of the theory of complexity applies in the case of life: „the whole is greater than the sum of its parts". (Kauffman[5] 1995, p. 15) Working within the unicorn world, this principle directly reflects the ontological contradiction created by human beings by breaking the Kant-Carnap rule and blending the EDWs. It is not possible to conceive of a whole and its parts within the same EW at the same time.

Applying the identity theory (the mind is the brain), it would be better to consider life identical with the organism/cell, even if this identity is also an erroneous notion.[6] Amazingly, nobody believes that life is identical with an organism or a cell, but many people believe that the mind is identical with the brain. Cognition/mind and life both have the same ontological status: they are both an EW, while neurons and other biological entities belong to the macro-EW. So, if life is in a similar position to cognition or mind, we cannot mix life with the corresponding cells or organisms. Moreover, we cannot mix the cells with matter that is, or belongs to, EDWs.

In this context, the life 'of an organism' is not the whole that is greater than the sum of its parts. Life merely corresponds to the sum of the cells and their activity. Translating 'the whole is greater than the sum of its parts', we get 'the whole corresponds to its parts'. I use inverted commas for its simply because no cell exists for its molecules, nor vice-versa. Imagine that you are a cell. Obviously, any cell (just like any other biological entity) has, as von Uexküll wrote, its own Umwelt. You, as a cell, are alive. You as a human being are alive, too. Both situations require the I/living being. The I/living being is the life that corresponds to the organism,

5 About Kauffman's theory of complexity and the EDWs perspective, see Vacariu and Vacariu (2010).

6 As I wrote in Vacariu (2008), in the unicorn world, the best alternative to the mind-body problem within the unicorn world was eliminative materialism. In one world (which has one ontology), epistemology and languages dissociations (which pretend to describe the phenomena of the same world from „various points of view") become useless. Even worse, the variations in vocabulary can produce a flawed framework which can lead scientists and philosophers to great, intemperate, and endless debates.

but, according to EDWs, a life does not exist/is for any other EW, i.e., an EW is not for any other EW.

I would like to draw your attention again to the fact that implicit knowledge is the I. What exactly does implicit knowledge mean for cells or insects? In the case of a cell, we may observe only the corresponding results of the implicit knowledge of a cell: the consequences of their biological functions. We have to change the notion of ontology regarding the relationships between microparticles-macroparticles, waves-corpuscles, and mind-brain. (Vacariu 2008) We have to change the notion of ontology regarding the existence of life for cells, insects or animals. It is much more difficult to identify the points of view of interactions/observations for living entities such as molecules, cells, neural patterns of activation, animals, and human organisms. Because of the I, any corresponding organism was able to survive in its environment and, in general, the evolution of species was possible.

The I (and even the organism) corresponds to an amalgam of biological or physical elements and their functions. We have to reject the continuity between the non-living matter and the living matter within the same world. Again, we must keep in mind that we need to reject the continuity between microparticles and macroparticles, between the organism and life, between the brain and the mind. Working only within the unicorn world until now, nobody has had any doubts about such continuities. By renouncing the world, we also renounce these continuities.

8
EDWs and Physics

8.1. Introduction

As I have shown in previous books, the unicorn world has been the wrong framework not only for philosophers, but also for scientists who do research in various domains (physics, biology, cognitive science). The scientists who work on fundamental scientific problems have to replace their current framework of thinking, the unicorn world, with EDWs. Throughout the history of human thought, few changes have been brought to the old framework. I hope that, after reading my books (2008, 2010, 2011, 2012, 2014), many people will find it clear that the unicorn world has to be replaced by EDWs. With my EDWs perspective, I offer the ontological (more precisely, the hyperontological) foundations of various dual phenomena (macro- and micro-phenomena, mental and neuronal phenomena, or life and organism/cell) investigated by people working in different fields (physics, cognitive neuroscience, and biology). In general, I applied my perspective to theories which many researchers acknowledge are highly problematic. This long-lasting research has not led to any answer precisely because these problems are actually pseudo-problems.

In my previous books, I have talked about two important errors that have appeared in physics: (1) in quantum mechanics there is a mixture of EDWs (the waves-EW and the microparticles-EW) (2) the attempt to elaborate the theory of everything, i.e. to unify quantum mechanics with Einstein's theory of relativity, has led to the mixture of EDWs (the macro-EW is mixed with the micro-EW). Until this year (2014), I have not applied my EDWs perspective to Einstein's theory of relativity simply because this theory is correct (both in its special and in its general forms). Even if the unicorn world was the dominant paradigm during his lifetime, Einstein created his theory for the macro-EW (or rather, a large number of macro-EWs, see below), and this is precisely why his theory is correct within the EDWs perspective.

In this chapter, I want to show that the EDWs perspective furnishes the (hyper)ontological foundations for Einstein's special and general theory of relativity, which have never been furnished by anybody. All researchers have worked within the unicorn world (including Einstein), therefore, it is no surprise that nobody was able to discover the foundations of the theory of relativity. First, I will deal with the hyperontological foundations of Einstein's special theory of relativity; then, I will analyze the general theory of relativity. I will do both from the EDWs perspective. In the second part of this chapter, I will show that quantum mechanics is a pseudo-theory constructed within the unicorn world by many famous physicists in the 20 century. In the last part, I will indicate that the search for the 'theory of everything' (the theory of unification) is a pseudo-investigation within the unicorn world due to the simple fact that the unification of Einstein's theory of relativity and of quantum mechanics represents a mixture of EDWs.

The EDWs replaced the unicorn world, the unique world, which had a unique ontology. Each EW has its own ontology, but since there are no multi-worlds, we cannot speak of multi-ontologies. One EW does not exist (it is not) for any other EW. Moreover, each EW has its own epistemological-ontology (the classical distinction between ontology and epistemology is, as I have explained, totally wrong). So, we can talk neither about ontological, nor about 'multi-ontological' (common to the multiverse), but only about hyperontological. This notion is abstract only because one EW does not exist for any other EW; instead, it refers to the sum of all EDWs in an abstract hyperontological framework analyzed by a human observer.

Both the special and the general theory of relativity involve many EDWs, therefore in this chapter I will introduce the hyperontological foundations of the special and general theory of relativity. Einstein's theory of relativity does not fit with our common sense, which is the result of the evolution of the species and the development of each individual in a standard environment. This standard environment refers, in this case, to existing at speeds much smaller than the speed of light.

> It has been almost a century since Einstein informed the world of his dramatic discovery, yet most of us still see space and time in absolute terms. Special relativity is not in our bones—we do not feel it. Its implications are not a central part of our intuition. (Greene 1999, p. 17)

Obviously, it has been difficult to fit the theory of relativity within what was generally considered to be reality (the world), when working within the paradigm of the unicorn world. The question is: how was it possible that, for a century, researchers understood that the absolute spatio-temporal framework does not exist, but instead there are only relative spatio-temporal frameworks (for instance, one from the point of view of a person in a train, and the other from the point of view of a person on the platform) if the framework was the unicorn world, a single ontological world? Let us suppose that the person A in the train and the person B on the platform outside the train both perceive a table on the ground. What do two people perceive? Do they perceive the same, identical table? Within the special theory of relativity, A and B exist in different spatio-temporal frameworks, therefore the table cannot be the same table. However, we cannot talk about two different tables, because in reality, in order to avoid an ontological contradiction, there are no absolute different tables. The correct answer can be given only within the EDWs perspective: these are epistemologically different tables in EDWs. This answer is banal, but it completely changes our thought framework.

In the past, nobody denied the existence of the world/universe/reality, and within this paradigm it was normal for the absolute spatio-temporal framework to be dominant. Einstein's theory of relativity rejects absolute space and time. Nevertheless, from my point of view, Einstein's theory of relativity has been formulated within the unicorn world, and our common sense reaches some ontological contradictions which simply cannot be accepted by our common minds. Adding to this the enormous difference between the speed of light and the speed of the objects surrounding us, we reach a situation in which, as Greene puts it, 'special relativity is not in our bones – we do not feel it'. We cannot observe the results indicated by the theory of relativity because these results are too small to be captured by our sensorial apparatus, which has evolved and developed within the macro-EW, in our standard environment. Newton's theory required absolute space, absolute time, and signals (the speed of light or gravitational waves) which can travel instantly. Until Einstein, nobody realized that the speed of light was the maximum possible speed, i.e. nothing can exceed the speed of light.

Even if Einstein's theory of relativity is perfectly correct, the EDWs perspective requires not only people working in cognitive (neuro)science, but also physicists, to change their framework of thinking. It implies that physicists could work not only with quantum mechanics, but also with the theory of relativity: they have to replace the unicorn world with the EDWs perspective in order to be capable of rejecting various pseudo-notions from quantum mechanics and to better understand (feel) the theory of relativity. The special and general theory of relativity can be applied to phenomena that belong to some EDWs, but they cannot be applied in every single case. For instance, these theories cannot be applied to the mind-EW, or to the micro-EWs.

Scientists have to understand that by working within a wrong framework, a pseudo-problem becomes a real problem, and they can waste their entire career creating extremely attractive Ptolemaic epicycles. Only by accepting the EDWs framework could a real progress in the particular sciences take place. Moreover, scientists can improve the scientific framework of EDWs.

In the following section, I will try to show that: (a) in some cases, the 'frames of reference' specified by these two theories are in fact EDWs; (b) the principle of relativity first proposed by Galileo, then redefined by Einstein ('The laws are the same for all observers in uniform motion'), is probably the strongest scientific argument to support the existence (more exactly, the being) of EDWs. This principle reveals that the world does not really exist, which therefore means that we have to replace the world with EDWs. (c) the EDWs perspective offers the most appropriate hyperontological foundations for both theories of relativity. Obviously, as a scientist, Einstein did not pay too much attention to the ontological foundations of his theory. Even if he completely changed the meaning of space and time, the notion of the world survived, i.e. neither Einstein, nor anybody else since then questioned the nature of the world. As opposed to the special theory of relativity, the general theory of relativity requires the existence of the universe more dramatically.

At the time when Einstein elaborated the special theory of relativity, many physicists were deeply concerned with absolute space and absolute time and light. At the end of the 19th century, there were two great theories in physics: Newton's mechanics and Maxwell's electromagnetism. In this context, as Greene emphasizes, two great problems arose: (1) The properties of the motion of light ('Can a beam of light be caught?' Newton's answer is affirmative, while Maxwell's answer is negative) (2) Nothing can travel faster than the speed of light (vs. Newton's gravity, which instantaneously influences any other objects) Later, in the first half of the 20th century, the third great problem appeared: (3) The incompatibility between Einstein's theory of relativity and quantum mechanics in explaining the world. (Greene 1999, p. 7)

8.2. EDWs and the special theory of relativity

Einstein created the special theory of relativity for phenomena that belong to frames of reference in uniform motion. The general theory of relativity is an extension of the special theory of relativity to non-inertial cases (in which acceleration and gravity are involved). In order to understand Einstein's special theory of relativity, we have to take into account two related aspects: (1) this theory concerns the motion of objects, as well as space and time (2) the movement of any physical object has to have four dimensions: three spatial and one temporal. The temporal dimension has two extreme points: (a) An object at rest moves with the speed of light in the fourth direction, that of time, therefore its temporal movement is at its maximum. (b) An object moving at the speed of light in three-dimensional space has no time dimension (the temporal dimension is zero), therefore its movement is shared (Greene's notion, see below) between three spatial dimensions. Einstein introduced two postulates:

(1) The principle of relativity: 'All frames of reference in uniform motion have the same physical laws'. This principle is applied not only to mechanics, but also to Maxwell's electromagnetic laws. I highlight that the principle of relativity greatly relates to one of my principles, i.e. the objective reality

principle: 'Any EW is, therefore all EDWs share the same objective reality, even if one EW does not exist for any other EDW'. (See the first chapter of this book)

(2) The second postulate is that „light is always propagated in empty space with a definite velocity c which is independent of the state of motion of the emitting body". Einstein gave no justification for this postulate in the introduction to his paper, so the strongest justification came from Maxwell's electrodynamics. That theory had equated „light with waves propagating in an electromagnetic field and concluded that just one speed was possible for them in empty space, c = 300,000 km/sec, no matter what the motion of the emitter". (Norton 2003, p. 5) Therefore, Maxwell's electromagnetic equations indicate that the speed of light is the same in all frames of reference and nothing can be faster than it.

Obviously, both postulates are essential for understanding Einstein's theory of relativity, but, at the end of the 19th century, these two principles seemed to be irreconcilable. Einstein solved this problem by changing the notion of simultaneity. (Norton 2003, p. 6) The apparent contradiction is the following:

> If one inertially moving observer measures c for the speed of some light beam, what must be measured by another inertially moving observer who chases after the light beam at high speed—say 50 percent of c or even 99 percent of c? That second observer must surely measure the light beam slowed. But if the light postulate respects the principle of relativity, then the light postulate must also hold for this second, inertially moving observer, who must still measure the same speed, c for the light beam. (Norton 2003, p. 6)

The relativity of simultaneity solved this apparent contradiction: „observers in relative motion do not agree on the simultaneity of events spatially separated in the direction of their relative motion". (Norton 2003, p. 6)[1] According to the special theory of relativity, not only the clock (and any process) in motion runs more slowly than a clock at rest, but also time itself runs more slowly for the clock.

To explain those classical two clocks: one clock is at rest, the other is in motion. For the clock at rest, time is given by the movement of a photon between two horizontal parallel mirrors at rest. The perpendicular movement of the photon from the first mirror to the second one and back represents time (for instance, one second). (Fig. 9a)

1 „The constancy of the speed of light requires that we give up the age-old notion that simultaneity is a universal concept that everyone, regardless of their state of motion, agrees upon. The universal clock previously envisioned to dispassionately tick off identical seconds here on earth and on Mars and on Jupiter and in the Andromeda galaxy and in each and every nook and cranny of the cosmos does not exist. On the contrary, observers in relative motion will not agree on which events occur at the same time." (Greene 1999, p. 21)

A second identical clock is in motion: as we have established, time is represented by the movement of a photon between those two mirrors in motion (one second per complete movement). The distance traveled by the photon (with the constant speed c) from the bottom mirror to the upper mirror and back is longer than the distance traveled by the photon moving between the same two mirrors at rest. Again, the movement of the photon between these mirrors represents a second. (Fig. 9b)

(a)

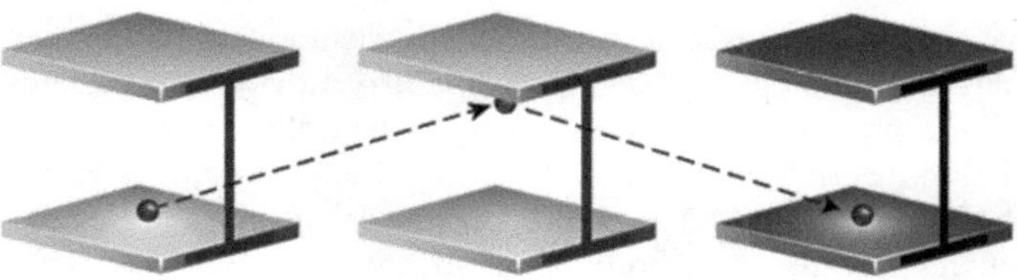

(b)

Fig. 9 Time = the movement of the particle
(Both illustrations are from Greene 1999)

In the second case, time runs slowly for objects (in our case, for the clock) in a moving state. If we increase the speed of the two mirrors, the distance covered by the photon will also increase and time will be dilated (it will run slowly). If those two mirrors reach the speed of light, the photon will never reach the second mirror. This means that time does not exist for entities (photons in this case) traveling with the speed of light. So, there are two frames of reference (the clock at rest and, respectively, the clock in motion), which differ in their temporal (and spatial) dimensions. Again, in motion, time runs slowly, therefore all processes run slowly as well, while space and the spatial dimension of objects are contracted (in the direction of motion). The degree of time dilation and space contraction depend on the speed of motion.

Again, within the EDWs perspective, this refers not only to spatio-temporal frameworks, but to EDWs, i.e. to the changes of some features pertaining to all objects/entities (and their relations) situated in that spatio-temporal framework. These spatio-temporal frameworks and all epistemologically different entities/processes are EDWs.

For a better understanding of this idea, I will use the EDWs perspective to analyze the principle of relativity by referring again to Einstein's famous train experiment (the simultaneity of two lights that strike the ends of the train): person A in a train has a constant speed in relation to person B on the platform, but is static in relation to the train and its objects; person B is at rest on the platform, but the train moves with a constant speed.[2] According to Einstein's special theory of relativity, motion modifies space and time, therefore it modifies the size of the objects (and their relations) in the direction of motion of the train in that particular spatio-temporal framework.

What is important for this thought experiment is that A and B are in 'different spatio-temporal frameworks'. But how can we understand the idea of 'different spatio-temporal frameworks' within the unicorn world? In the unicorn-world, can the answer be that we are talking merely about spatio-temporal frameworks epistemologically, not ontologically? The principle of relativity demonstrates that neither A, nor B can proclaim themselves as being in the real state of rest and in the real state of motion, respectively. That is, A can say 'I am at rest, B is in motion', while B can think 'I am at rest, A is in motion'. As Wolfson (2000) emphasizes several times in his documentary, both A and B are right. Einstein solved this apparent contradiction by changing the notion of simultaneity (space and time): two events that are simultaneous for A cannot be simultaneous for B. According to Einstein, A and B are in different spatio-temporal frameworks. If we do not agree to this, we have to give up either the principle of relativity (and to work within the absolute frame of reference, i.e. the absolute spatio-temporal framework), or the constancy of the speed of light. Again, Einstein avoided the incompatibility between the principle of relativity and the constancy of the speed of light by changing the notions of space and time.

Simultaneity exists in two 'spatio-temporal frameworks'. Again, what exactly does framework mean in this context? Within the unicorn world, these two spatio-temporal frameworks produce an ontological contradiction. From the EDWs

2 In the past, A and B needed to both be at rest on Earth. Then A starts to accelerate until reaching a particular constant speed in relation to B and the Earth. Nevertheless, the processes of acceleration and deceleration are explained by the general theory of relativity, which is analyzed in the second part of this work.

perspective, we have to reinterpret the notion of simultaneity, an essential phenomenon for the special theory of relativity. These spatio-temporal frameworks are in fact EDWs. Essentially, according to the 5 Principle, one EW does not exist for any other EW, even if we can indeed check for a very approximate correspondence (simultaneity).

The EDWs perspective furnishes the hyperontological foundations for Einstein's simultaneity, which is based on the fact that the speed of light is not infinite, but it is the maximum speed that can be reached. Essentially, two events that are simultaneous in one spatio-temporal framework cannot be simultaneous in another spatio-temporal framework. We have to remember Einstein's verdict: each frame of reference has its own time (Einstein 2011, section 8). For instance, from B's point of view A is moving at a constant speed (A has a rectilinear, uniform movement), while B is at rest (in relation to A). However, A can consider that B is moving, while A is at rest. Who is right? Obviously, according to the principle of relativity, both are right. Within the unicorn world, this type of judgments (which involve the principle of relativity elaborated by Galileo) produce (hyper)ontological contradictions.

With the aid of the EDWs perspective, we can avoid these contradictions: A and B are in EDWs and, more importantly in this context, the EW of A does not exist for the EW of B, nor vice-versa. In the A-EW, we can find A and the image of B; in the B-EW, we can find B and the image of the A. Only within the EDWs perspective, are 'both A and B correct'. Thus, only with the EDWs perspective could we extend the main idea of Einstein's special theory of relativity from 'different spatio-temporal frameworks' to EDWs.[3]

Another important aspect of Einstein's special theory of relativity: space and time are not separate features of the world. Since Einstein, we have come to know that space and time cannot be separated (in fact, Minkowski elaborated the mathematical framework for the unification of space and time in the spatio-temporal framework[4]). Therefore, we can only talk of the spatio-temporal framework, not of space or time in general.

As we have seen above, this paradigm shift changes not only the spatio-temporal framework, but also the size of the objects (in the direction of motion) in that particular spatio-temporal framework. Therefore, regarding the relationship

3 Moreover, as I have shown in my previous books, we must also include the I, the objects and the perceptions of objects in this analysis. The perceptions are the I, which is an EW, while the objects belong to an EDW.

4 „The views of space and time which I wish to lay before you have sprung from the soil of experimental physics, and therein lies their strength. They are radical. Henceforth space by itself, and time by itself, are doomed to fade away into mere shadows" (Kaku 1994, p. 85). Also, „Time being the fourth dimension means that time is intrinsically linked with movement in space. How fast a clock ticks depends on how fast it is moving in space". (idem)

between the spatio-temporal framework and any of the objects in it, we have to apply the same principle of unification: from my point of view, we cannot talk of the changes of that spatio-temporal framework (space contraction, time dilation), but only of the changes brought to an EW as a whole (including to its entities and their relationships). Obviously, Einstein also refers to the changes brought to entities (and to their relationships), but the problem appears when we notice the ontological contradictions between the two spatio-temporal frameworks of A (in the train) and B (on the platform) within the same world. Again, in order to avoid such hyperontological contradictions, we have to replace the unicorn world with the EDWs perspective.

This is another important reason for us to let go of the unicorn world and accept the existence of EDWs. The principle of relativity indicates that each observer can consider himself/herself in a state of rest, and the other person in a state of motion. However, even if we accepted that space and time really exist, we would have to translate the special theory of relativity in terms of the EDWs perspective: two observers who are in uniform motion – in relation to each other – are in EDWs. Therefore, we must replace spatio-temporal frameworks with EDWs. We have to remember that, within the EDWs perspective, the distinction between ontology and epistemology does not exist. It is precisely this status (which is tightly related to the principle of relativity) that forces us to move from different 'spaces-times' to EDWs.

Both from Leibniz's point of view and mine, space and time have to be correlated with something that belongs to/characterizes/refers to an EW. Regardless of whether space and time have an ontological substance or not (whether they exist or not), the epistemologically different entities (which have features such as being in a state of rest or in motion in a particular frame of reference) and their relationships really belong to the EDWs. We might doubt the ontological status of space and time, but we cannot deny the existence of any real object and its movement.

We do not actually perceive space or time, but we do perceive the motion of various entities in different 'frames of reference'. In fact, our perception is not restricted to the motion of the entities: in reality, we perceive a scene (which can include the motion of an object) that belongs to a particular EW. Therefore I conclude that A and B are in EDWs. Obviously, Einstein could not use EDWs instead of different spatio-temporal frameworks. From my point of view, it is not only that those pairs of events are not simultaneous, but, moreover, one pair of events does not exist for the other pair of events, since they belong to EDWs. Two events that take place simultaneously in the A-EW (which includes A and B from the A-EW) only correspond to two non-simultaneous events in the B-EW (which includes A and B from the B-EW). As we already know, it is not only wrong, but meaningless to think that the epistemologically different events which involve epistemologically different entities and/or processes situated in EDWs 'occur in the same spatio-temporal framework'.

Despite working in the unicorn world, Einstein constructed the theory of relativity, which is correct for macro-EDWs.[5] However, he could only talk of different spatio-temporal frameworks without any ontological support (since all such frameworks are within the same unicorn world).[6] This pair of events is neither identical (which would require the existence of the unicorn world), nor different (which would require a kind of Cartesian dualism for the world, i.e., the multiverse, for instance, or parallel worlds). In order to avoid both the unicorn world and the multiverse, we should return to the last part of the 5 Principle: one EW does not exist for any other EW. Einstein, as well as all the other researchers, could have talked of simultaneity working within the unicorn world and projecting a representation of each phenomenon in every mind (each mind being an EW). In this way, we can talk of the simultaneity of two events within the human mind, but externally, each event belongs to an EDW.

I will now look at Einstein's spatio-temporal four-dimensional framework in more detail. By analogy with the idea of rotation in a three dimensional space, we can imagine the rotation of an object within the four-dimensional framework. (Kaku 1994) That is, the motion of the object in the temporal dimension can be transformed into the motion of the object in a spatial dimension, or vice-versa. Motion is the key element in this situation in particular, but also in Einstein's theory of relativity in general. Motion is shared between these four dimensions, three spatial and one temporal.

> Einstein found that precisely this idea—the sharing of motion between different dimensions—underlies all of the remarkable physics of special relativity, so long as we realize that not only can spatial dimensions share an object's motion, but the time dimension can share this motion as well. In fact, in the majority of circumstances, most of an object's motion is through time, not space. (Greene 1999, p. 27)

In Einstein's view, all objects always travel with the speed of light through the spatio-temporal framework. (Greene 2011, p. 67) If an object is at rest spatially, the entire motion happens in the temporal dimension, moving the object in time. (We have to remember that each frame of reference has its own time. Einstein 2011, section 9) As Greene explains, if a particle is at rest relative to a particular four-dimensional frame of reference (three dimensions are spatial, one is temporal), the

5 We cannot apply Einstein's theory of general relativity in all EDWs: for instance, we cannot apply this theory to phenomena that belong to the mind-EW.

6 With the principle of relativity, Einstein should have rejected the existence of the world: if everything is relative, absolute space and absolute time do not exist, hence the absolute world (the unicorn world) does not exist, either!

entire motion of that object takes place in the temporal dimension. If the object moves in space, then a part of the object's temporal motion has to be converted into spatial motion. (Greene 2003, p. 27) Therefore, if an object moves through space, a part of its motion through time is deviated into its motion through space. It means that the object moves slower in time, i.e. its clocks move slower. (idem)

> We now see that time slows down when an object moves relative to us because this diverts some of its motion through time into motion through space. The speed of an object through space is thus merely a reflection of how much of its motion through time is diverted. (Greene 2003, p. 27)

The maximum speed in the spatial dimension can only be reached when the entire motion through time is diverted into motion through space. This is why time does not exist for photons. „This occurs when all of its previous light-speed motion through time is diverted to light-speed motion through space. But having used up all of its motion through time, this is the fastest speed through space that the object—any object—can possibly achieve." (Greene 2003, p. 27) From my point of view, motion determines the spatio-temporal framework, and this framework is an EW. We have to be aware of the fact that motion is not the only feature that determines an EW. However, we have to apply this parameter (motion, i.e. Einstein's theory of relativity) to any particular EW that has a spatio-temporal framework. Nonetheless, in this type of an analysis, we have to combine all the parameters which determine a particular EW – for instance, we have to combine motion with size.

If time does not exist for photons (for us as their observers, time does exist, because we are in a different frame of reference), then does space exist for photons? If yes, what kind of space? To answer this question, we must keep in mind an important detail concerning photons: they do not have mass. If photons have zero mass and time does not exist for them, then how does space exist for these micro-particles? Perhaps photons move in a space that only exists for their external observers (who are, anyway, in a different frame of reference, i.e. EDWs). The principle of relativity (or the principle of objective reality in my EDWs) leads us towards the following conclusion: the interaction between a photon and any other entity takes place in at least two EDWs: in the EW of photons and in the EW of the other entity. Generally, the interaction between two entities which belong (for instance) to two EDWs happens in two EDWs.

In this context, Einstein's famous thought experiment (in which he is riding on a photon, questioning the speed of other photons) has to be completed with the details provided by the EDWs perspective. Let us imagine that A is situated on top of a photon (which moves at the speed of light) and B is in a state of rest on Earth. According to the special theory of relativity, light has the same speed c for both A and B (who are in different frames of reference or EDWs). For A, time does not exist. The problem is that A has a mass (any kind of observer has a mass), therefore

A cannot reach the speed of light even in a thought experiment (A would require an infinite amount of energy to reach the speed of light).[7] Therefore, from this perspective, Einstein's thought experiment is impossible. Only a zero mass entity could reach the speed of light. A photon cannot see/perceive in the real sense (but, from my point of view, a photon interacts with its environment) simply because, from what we know and as far as we can conceptualize, any kind of perception would require a mechanism with a particular mass.

From all of the above, I conclude that Einstein's special theory of relativity is a scientific argument which clearly indicates the existence (or, more exactly, the being) of EDWs: if absolute space and absolute time do not exist, then indeed there is no world/universe. On the contrary, even if there are notions such as the world or the universe or reality, which were created by human beings in the dawn of our history, the world in fact does not exist. However, this does not mean that we have to take refuge in Berkeley's absolute idealistic framework (the world and matter do not exist, therefore everything is spiritual or God), not even if an EW has no spatio-temporal framework. Those who accept Berkeley's view should go to church, not work in the academic world.

The unicorn world has been the direct consequence of two wrong ideas which have dominated the thought framework over the millennia: (1) the unquestioned existence of a spatio-temporal framework shared by all phenomena (the same world/universe/reality external to the perceptions of the I) (2) the distinction between ontology and epistemology.

However, Einstein's theory of relativity, as well as quantum mechanics (non-locality and non-spatiality, for instance) compel us to rethink our concepts of space and time and therefore to question the ideas we have concerning the world/the universe. Einstein's theory has led to the rejection of absolute space and time, but, working within the unicorn world, we reached the very serious (hyper)ontological contradictions mentioned above. Moreover, in quantum mechanics there are some essential problems (non-locality, for instance – see below), which do not yet have an alternative, and which I believe to be in fact pseudo-problems constructed within the unicorn world, since we have reached some hyperontological contradictions in this area, as well. (See Vacariu 2008, Vacariu and Vacariu 2010).

That all entities and their relationships belong to EDWs has to become much more palpable. If we consider the size-parameter, it becomes clear that all human bodies (and their brains) are in the same macro-EW, but each mind (life/the I) is an EW. However, in this book, working on the relationship between Einstein's theory of relativity and EDWs, we can see that the motion-parameter hints to the existence of very numerous, uncountable macro-EDWs. In order to describe a particular EW,

7 A better framework for Einstein's thought experiment would be if Einstein himself were a photon.

we have to take into account all the parameters which characterize EWs in general. As I have shown in this book, the characteristics of EWs can differ greatly – for example, space (or color) does not exist in the mind-EW; I consider that human beings have only the idea of space, we do not have any kind of spatial representation.

8.3. EDWs and the general theory of relativity

The EDWs perspective furnishes the hyperontological foundations for Einstein's general theory of relativity and, consequently, it furnishes the foundations for the special theory of relativity as well, since the latter is merely a particular case of general relativity. The special theory of relativity refers to the frames of reference when the system has a uniform motion (inertial reference frame) or when it is at rest, while the general theory of relativity refers to all states, including accelerating frames of reference, or gravitational fields (in other words, non-inertial frames).

General relativity relies on a general principle of relativity: 'All bodies of reference K, K', etc., are equivalent for the description of natural phenomena (formulation of the general laws of nature), whatever may be their state of motion.' If Einstein's special theory of relativity replaced Newton's classical mechanics, then the general theory of relativity replaced Newton's theory of gravity. To start with, Newton's theory of gravitation was inconsistent with special relativity. Since Newton's equation of gravity lacked a time component, it was possible to think that gravity acts instantly at great distances. This possibility was against the special theory of relativity (nothing can be faster than the speed of light, c). Moreover, in 1907, still working in the patent office (where he elaborated special relativity), Einstein wanted to extend his special theory (which referred to inertial frames of reference) to all possible phenomena.

> My first thought on the general theory of relativity was conceived two years later, in 1907. The idea occurred suddenly. I was dissatisfied with the special theory of relativity, since the theory was restricted to frames of reference moving with constant velocity relative to each other and could not be applied to the general motion of a reference frame. (Einstein 1982, pp. 46-7)

In order to develop special relativity, Einstein came up with a thought experiment: 'What would I perceive if I were riding a photon?' For general relativity, he had another thought experiment: a man falling off the roof would not feel his weight (this was, he said, the 'happiest thought of his life').

> The breakthrough came suddenly one day. I was sitting on a chair in my patent office in Bern. Suddenly a thought struck me: If a man falls freely, he would not feel his weight. I was taken aback. This simple thought experiment made a deep impression on me. This led me to the theory

of gravity. I continued my thought: A falling man is accelerated. Then what he feels and judges is happening in the accelerated frame of reference. (Einstein 1982, p. 47)

This falling man,

does not feel his weight because in his reference frame there is a new gravitational field which cancels the gravitational field due to the Earth. In the accelerated frame of reference, we need a new gravitational field. (Einstein 1982, p. 47)

In this way, Einstein moved from the particular case of inertial motion to general cases of motion (including accelerated motion and gravitation). In this context, Einstein needed to replace Euclidean geometry with non-Euclidean geometry and to introduce the principle of equivalence between acceleration and gravitation which was illustrated by the thought experiments below. There are two cases of equivalence:

(1) Imagine a man standing in an elevator near Earth. Somebody cuts the elevator's cables, which leads to the man and the elevator falling freely due to the gravitational field. However, because of the acceleration, the man in the elevator would not feel his weight, he would be weightless. As Wolfson emphasizes (2000, episode no. 13), it does not mean that there is no gravity in the elevator's area. The elevator is in a field of gravity, but another force (given by the accelerating movement) annihilates the force of gravity for the man in the elevator. The principle of this annihilation is based on the identity between the gravitational mass and the inertial mass. This situation is equivalent to one in which the elevator and the man are in uniform motion in intergalactic space. What is important to remember in this case is that Einstein realized gravity does not exist as a force, since in these cases it disappears.

(2) If a man is in an elevator in deep space (where there is no gravity) and someone gives the elevator a certain acceleration, the man in the elevator would have the impression that he could feel a gravitational field. Similarly, if the elevator and the man are on Earth, Earth's gravity acts on both of them.[8] In this case, the action of gravity and the action of acceleration are indistinguishable. (Wolfson 2000)

8 Wolfson observes that, for Einstein, simultaneous events are in one frame of reference, but not in another, therefore time has no objective reality (the spatio-temporal framework becomes relative); similarly, in general relativity, if acceleration exists in one frame of reference, but in another frame it is gravity, not acceleration, then acceleration/gravitation are relative – more precisely, gravity does not exist. There has to be something that replaces gravity, something that is constant and has a continuity. According to Einstein, this has to be not another force, but the geometry of space-time, i.e. the four-dimensional non-Euclidean geometry. (Wolfson 2000, episode 13) This idea is related to Riemann's idea of force as equivalent to geometry (see later the footnote about this notification.)

This thought experiment mirrors the principle of equivalence between accelerated motion and gravity, or the identity between gravitational mass and inertial mass.[9] Wolfson indicates two important statements for general relativity: (1) Matter and energy curve spacetime. (2) Objects move in the straightest possible paths in curved spacetime. (Wolfson 2000, episode 14) So, Earth (just like all macro-objects) curves the surrounding space (and time), and this is why all objects near our planet have a tendency to fall on it. Space itself is curved and time runs more slowly near a planet than in empty space. Again, we are dealing with the deformation of the spatio-temporal framework: mass curves spacetime. For instance, identical clocks at different altitudes will run at different speeds, or identical clocks with different accelerations will similarly run at different speeds.

I will now analyze, from my point of view, the case of the man in the continuously accelerating elevator. In this case, in what EW is the observer/entity? The answer is that the observer/entity passes through uncountable EDWs, determined by the observer's dimensions and acceleration. But taking into account only the acceleration (not the difference between the mind and the brain, or other parameters related to the abilities of perception) it is not the same observer moving through EDWs, but there are epistemologically different observers in EDWs. From this example, we can observe that even if a parameter which determines EDWs (the rates of change for the speed/motion in this case, i.e., whether the vehicle accelerates or decelerates) seems to change continuously, this does not imply an infinity of EDWs, but only uncountable EDWs. There are other parameters which can change their values in uncountable time: for instance, the size of entities or/and the gravity (i.e. the deformation of space and time).

We return to the example in which person A is in the train and person B is on the platform. Moving with a constant speed, A does not perceive/feel anything regarding his/her constant and uniform motion (the principle of relativity). However, if the value of the speed changes (the train accelerates or decelerates), A can perceive the change. If the train transporting A is decelerating and eventually stops (in relationship with B), then A (moving through an uncountable number of EDWs) will be (in the end, i.e. when A stops in relationship to B) in B-EW, which is the macro-EW (i. e. in that EW formed by all the macro-objects that are on Earth). As a consequence of the fact that the A-EDWs are created because of his/her motion, these EDWs simply disappear into hypernothing.

9 For Einstein, this equivalence was given by the equivalence between the inertial mass and the gravitational mass. As he mentions in his 1916 book, people working in mechanics knew about this identity, but they never interpreted it. The identity between the inertial mass and the gravitational mass is the postulate of the general theory of relativity. „The laws of nature in an accelerating frame are equivalent to the laws in a gravitational field". (Kaku 1994, p. 89, italics in the original)

According to the general theory of relativity, the curvature of space and time is gravitation, that is, the presence of mass (for instance, the presence of the sun) warps the space that surrounds it (which remains flat in the absence of massive objects) (Fig. 10) and time runs more slowly in the gravitational field than in empty space, where there is no gravitation. Gravity is replaced by this warped space.

In order to better understand this, we can recall the following thought experiment: a beam of light enters an accelerating elevator through a hole in the left wall. For the person inside the elevator, the beam of light will strike the right wall in a point closer to the floor. Einstein noticed that the light's path is curved, which led him to believe that, if acceleration bends the path of the light, then according to the principle of equivalence, gravitation similarly curves the light-path. More precisely, massive objects (planets, galaxies) warp space (it has been said that gravitation does not act on photons, since photons have no mass, but in reality photons follow the curves created by planets or galaxies when they deform space).

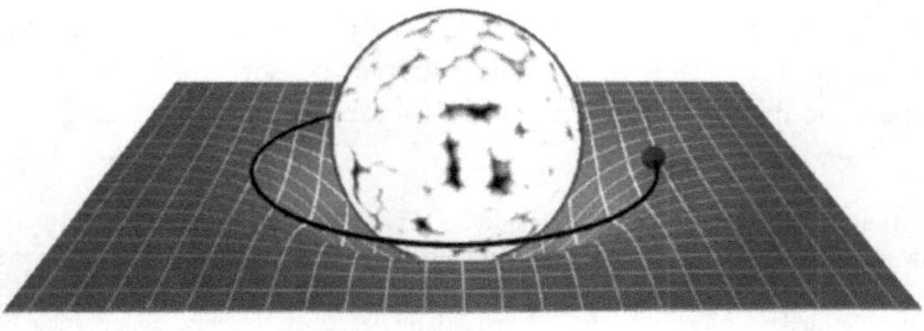

Fig. 10 There is no gravitation, only the distortion of space-time by massive objects (planets)
Gravitation does not exist. The sun warped space (time) around it. In this drawing, for the sake of convenience, the distortion is illustrated only on the horizontal plane. In reality, however, the sun warps space in every direction. The Earth is kept on its orbit because of this warped space.

8.4. EDWs and quantum mechanics[10]

If we consider that Max Born was completely right („I am now convinced that theoretical physics is actual philosophy", one of the mottoes of this book), I believe that the EDWs perspective could be the best alternative for explaining the most important (but problematic) notions/phenomena of quantum mechanics: Bohr's

10 For more details regarding EDWs and quantum mechanics, see the last chapter of Vacariu (2008) and Vacariu and Vacariu (2010).

complementarity and superposition, entanglement, non-locality and non-separability. In this section, I propose a new interpretation of quantum mechanics, while at the same time changing the old thought framework of quantum mechanics (the world, i.e., the unicorn world, the framework which produces the weirdness of this theory) with a new one, the EDWs perspective.

The Copenhagen interpretation (proposed by Bohr and his team) is one interpretation of quantum mechanics, but it makes the same error of assuming the existence of the unicorn-world, as we can see from its approach to the problem of measurement.[11] In the Copenhagen interpretation, a subject can observe the wave at one moment, using one tool of observation. When she changes the measurement apparatus in order to observe an electron, the wave function collapses at a certain location. This collapse is produced by the measurement apparatus. Bohr always emphasized that it is meaningless to ask about the position of an electron before it is measured. For Bohr, „the electron simply does not have a definite position before the measurement is taken". (Greene 2004, p. 94)[12] The error that made its way into this concept is that three objects are postulated in the same unicorn-world – the wave that collapses, the electron (a microscopic object) and the measuring instrument (a macroscopic object).[13] To avoid this paradox, Bohr's stratagem was to

11 „In his solar-system atom model (Figure 7.5, middle), Niels Bohr changed the second part of the description by introducing quantum jumps between special orbits, but he kept the first part. Schrödinger was even more radical, and changed the first part too: he abandoned the very idea that a particle has a well-defined position and velocity! Instead, he described the state of a particle by a new mathematical beast called a wavefunction, written ψ, which describes the extent to which the particle is in different places. Figure 7.5 (right) shows the square2 of the wavefunction, |ψ|2, for the electron in a hydrogen atom in an n = 3 orbit, and you can see that rather than being in a particular place, it seems to be on all sides of the proton equally, while preferring certain radii over others. How intense the 'electron cloud' of Figure 7.5 (right) is in different places corresponds to the extent to which the electron is in these places. Specifically, if you experimentally go looking for the electron, you find that the square of the wavefunction gives the probability that you'll find it in different places, so some physicists like to think of the wavefunction as describing a probability cloud or probability wave. In particular, you'll never find a particle in places where its wavefunction equals zero. If you want to stir up a cocktail party by sounding like a quantum physicist, another buzzword you'll need to drop is superposition: a particle that's both here and there at once is said to be in a superposition of here and there, and its wavefunction describes all there is to know about this superposition." (Tegmark 2014, p. 160)

12 „When Bohr exclaimed, 'No reality without observation!' it seemed to put humans back on center stage. After Copernicus, Darwin and others had gradually deflated our human hubris and warned against our egocentric tendencies to assume that everything revolved around us, the Copenhagen interpretation made it seem as if we humans in some sense created reality by just looking at it." (Tegmark 2014, p. 163) The EDWs perspective is totally contrary to Bohr's view: we, humans, do not create reality! However, the EDWs perspective follows Copernicus, Darwin, Freud and Einstein's line of thinking: even the world/universe does not exist, but the self is an EW.

13 „Essential to the Copenhagen interpretation was a clear separation of the quantum microworld from the classical macroworld. That separation depended on a vast difference in scale between atoms and

negate the existence of the particle until that particle is observed, at which moment the wave function collapses into the electron at a certain location.[14]

Bohr's approach represents one extreme position. The other extreme position concerning the quantum measurement problem is the many-worlds approach (Everett, De Witt, Deutsch, etc. – see below). Between these extremes there are other approaches, but all these theories assume the existence of the unicorn-world.[15] To offer a better general image of quantum mechanics, I will first introduce a few ideas about the problems from quantum mechanics. Then, I will analyze these notions within the EDWs perspective. Finally, I will end this section by analyzing recent papers written by some physicists on the same problematic notions of quantum mechanics. I am directly interested in analyzing how scientists constructed their alternatives as they tried to solve quantum mysteries within the unicorn-world paradigm.

Taking collapse as a criterion, Putnam classifies the different interpretations of quantum mechanics. (Putnam 2005) He replaces the measurement problem from quantum mechanics with the collapse problem. For him, the question is „Do we or don't we need to postulate a collapse and if we do assume a collapse, what should

the things we deal with directly. In Bohr's day, there was a wide no-man's land in between. It seemed acceptable to think of the macro realm obeying classical physics and the micro realm obeying quantum physics." (Rosenblum and Kuttner 2006, p. 111) From my viewpoint, it is indeed a separation of microworld from macroworld, but these worlds are EDWs, and more important, one EW does not exist for any other EDW. Only with the EDWs perspective can we reject this „wide no-man's land"!

14 „I think that a particle must have a separate reality independent of the measurements. That is, an electron has spin, location and so forth even when it is not being measured. I like to think the moon is there even if I am not looking at it." (Einstein in Rosenblum and Kuttner 2006, p. 125) „At the 1927 Solvay conference, Einstein, by then the world's most respected scientist, turned thumbs down on the newly minted Copenhagen interpretation. He insisted that even little things have reality, whether or not anyone is looking. And if quantum theory said otherwise, it had to be wrong. Niels Bohr, the Copenhagen interpretation's principal architect, rose to its defense." (Rosenblum and Kuttner 2006, p. 125) From my point of view, the EDWs perspective, a particle has a „separate reality independent of measurement". That particle exists because of its interactions with other microparticles from the same EW.

15 There are several approaches to the quantum measurements problem, the main ones being: the Copenhagen interpretation (headed by Bohr), the many-worlds approach (Everett, Deutch, etc.), Bohm's approach, and Girardi, Rimini & Weber's approach. (See Putnam 2005; Greene 2004) In Chapter 14, Rosenblum and Kuttner (2006) write about nine interpretations of quantum mechanics. Trying to explain the phenomena (the empirical measurements), different researchers introduced Ptolemaic epicycles in constructing various alternatives to the quantum mechanical-world, but because they work within the unicorn-world, their approaches are wrong. We know from the history of human thought that human imagination has played a powerful role in creating ardent arguments for fanciful Ptolemaic epicycles.

we say about it?"[16] (Putnam 2005, p. 624) The collapse problem refers to the relationship between waves and microparticles, or between micro- and macro-observables.

From the EDWs perspective, we can see this kind of interaction as a Ptolemaic epicycle of the unicorn world, and the interpretations of quantum mechanics as paradigmatic Ptolemaic epicycles arising from a pseudo-problem. Putnam is right in saying that the interpretation of quantum mechanics is 'a philosophical problem in detail', but he is wrong when he claims that scientific realism is the premise of the discussion. Scientific realism refers to the scientific unicorn-world. The rejection of the unicorn-world can be done only from a philosophical meta-paradigm, and this is the EDWs perspective. As we will see in this section, all the authors mentioned below offer alternatives to the same problems. After presenting various interpretations of quantum mechanics, using collapse as a criterion, Putnam classifies the main alternatives: the Von Neumann, many-worlds, Bohm and Ghirardi-Rimini-Weber interpretations. As I have mentioned above, one of the main problems in quantum mechanics is the relationship between micro- and macro-particles. Putnam reconsiders the Copenhagen interpretation in the following light:

> [T]hat macro-observables retain sharp values at all times […], while micro-observables have sharp values only when measured, where measurement is to be defined as a certain kind of interaction between a micro-observable and a macro-observable. (Putnam 1965, pp. 149–55 in 2005, pp. 624–5)

The remaining problem for quantum mechanics (Putnam 1965, p. 157) was therefore to figure out what makes the macro-observable special: „The result we wish is that although micro-observables do not necessarily have definite numerical values at all times, macro-observables do". (Putnam 2005, p. 625)

Later, Putnam wrote that the first alternative in his classification (Von Neumann – the collapse is produced by an external factor) presupposes that the collapse is something external to the system and not subject to superposition, which is an unsolved problem.

16 Everybody knows that quantum mechanics has incredible empirical results in its application. However, Greene mentions that „After more than seven decades, no one understands how or even whether the collapse of a probability wave really happens". (Greene 2004, p. 119; italics in the original!) Or Davies: „Although quantum mechanics is a breathtakingly successful theory in its application, its interpretation remains confused and hotly debated." (Davies 2006, p. 290) Evidently, within the unicorn-world, nobody could explain what happens with the wave during the measurement of the position of an electron.

> „Macro-observable is not the sort of term that can be an irreducible primitive in an ultimate physical theory, so I called for some future extension of quantum mechanics that would explain why macro-observables do not go into such states as $1/\sqrt{2}$(Live Cat) + $1/\sqrt{2}$(Dead Cat)". (Putnam, p. 628)

Let us analyze this paragraph in more detail. I believe that the framework of the unicorn world forces scientists to introduce the 'collapse of the wave'. Otherwise, they could not accept that two things simultaneously exist in the same place, at the same time, or that the same thing has two positions (one of which being observed or measured by us) within the unicorn world. In other words, planets, for instance, could not be irreducible primitives. Therefore, scientists have been struggling to find the gravitons which cause of gravity. (Gravitons will be discussed further below.)

From my point of view, planets are indeed irreducible primitives because, according to Einstein's general theory of relativity, gravity is caused by massive objects that warp the surrounding space, gravity being a property of space. As we saw above, ignoring microphysical forces, Einstein adopted a perspective on the relationship between this necessary geometry and entities as „practically rigid bodies". (Friedman 2001, p. 114) Within EDWs, we do not ignore any forces and each EW has its own irreducible primitives. Waves, planets and macro-objects are the irreducible primitives of the macro-EW and microparticles are the irreducible primitives of the micro-EW. The 'theory of everything' is the 'theory of the unicorn world'. (Superposition and Putnam's many-worlds interpretation are discussed below.)

Putnam concludes that are three alternatives: Bohm's or Ghirardi-Rimini-Weber's, which is the correct interpretation, Pitowski's alternative (Pitowski comments on one of Putnam's conferences, as Putnam himself remarks), which indicates that we will „fail to find a scientific realist interpretation which is acceptable". (p. 631) (As we will see below, Bohm's interpretation is the closest alternative to the EDWs perspective. However, his alternative is also constructed within the unicorn world). From an EDWs perspective, it is indeed impossible to find a scientific realist interpretation of quantum mechanics within the unicorn world. Replying to Putnam's presentation, Pitowski said „You are saying that before we can interpret quantum mechanics we have to change it". (Putnam 2005, p. 632) Putnam's answer was that Von Neumann

> [...] already changed quantum mechanics, certainly from Bohr's point of view. All interpretations of quantum mechanics are in a sense changes of quantum mechanics, because it is an incomplete theory—one cannot regiment it, formalize it in standard logical notation [...] unless you add an interpretation. (Putnam 2005, p. 632)

From my perspective, any interpretation of quantum mechanics is an incomplete theory within the unicorn world, and the search for hidden variables

(Ptolemaic epicycles) is useless. Using different macro tools of observation, a human being can observe, at different times, the electron and the wave belonging to EDWs. This idea can become clearer if we imagine a girl who was born without any of the normal senses, but with an electron microscope instead of eyes. For the 'girl with the electron microscope instead of eyes', microparticles and their relationships exist, but tables, electron microscopes and planets do not. The much-wanted interaction between micro- and macro-particles exists only in the unicorn world.

It is an error to consider that the wave, the electron, and the macro tool of observation are in the same unicorn world. The wave and the electron both exist at the same time, but in EDWs. In fact, the electron from one EW corresponds to the wave from another EW.[17] The collapse of the wave represents the switch made by the observer between one EW and another through the use of different tools of observation. The EDWs perspective offers a simple explanation of the infamous property of non-locality.

For instance, let us consider how we can measure the spin or polarization of two particles which belong to EW1. These particles, which initially represent one system, are later separated. According to the Copenhagen interpretation, the spin of particle 1 has no value until it is measured. Before the measurement, there is a superposition of various states of that particle produced by the unitary evolution of the wave function which corresponds to that particle. The act of observation produces the collapse of the wave function, and this means that the observer sees the particle in one definite classical state. The measurement of the spin of the first particle (for example, the up state), which produced a collapse of the wave function, has an instantaneous effect on the spin of the second particle (the down state).

17 „In the 1920s, Schrodinger and Heisenberg independently demonstrated that it was possible to derive these rules from first principles if electrons obeyed rules of dynamics that were different from those applied to macroscopic objects like baseballs. Electrons could behave like waves as well as particles, appearing to spread out over space (hence, Schrodinger's wave function for electrons), and the results of measurements of the properties of electrons were shown to yield only probabilistic determinations, with various combinations of different properties not being exactly measurable at the same time (hence, Heisenberg's Uncertainty Principle)." (Krauss 2012, pp. 74-5) Electrons do not „behave like waves as well as particles", they are instead particles which correspond to the waves (that belong to an EDW). „In summary, Schrödinger altered the classical description of the world in two ways: 1. The state is described not by positions and velocities of the particles, but by a wavefunction. 2. The change of this state over time is described not by Newton's or Einstein's laws, but by the Schrödinger equation. These discoveries by Schrödinger have been universally celebrated as among the most important achievements of the twentieth century, and they created a revolution in both physics and chemistry. But they also left people tearing their hair out in confusion: if things could be in several places at once, why did we never observe that (while sober)? This puzzle became known as the measurement problem (in physics, measurement and observation are synonyms)." (Tegmark 2014, p. 162) The electron is not in „several places at once", it is only in one place as we observe it, but it corresponds to the wave from an EDW. The puzzle of 'measurement problem' is solved with the EDWs perspective!

Under the Copenhagen interpretation, this instantaneous effect represents action-at-a-distance, or faster than light transmission[18], which Einstein's special theory of relativity tells us is not possible.

Einstein and his colleagues claimed that quantum mechanics is incomplete because it does not take into account certain hidden variables of reality. On the other hand, Bell's inequality assumes Einstein's condition of locality is true.[19] The experiments which involve the measurement of correlated photons[20] (with a

18 „In refuting EPR, Bohr claimed that what happened to one object could indeed influence the behavior of the other instantaneously, even though no physical force connected them. Einstein derided Bohr's influences as 'spooky interactions'." (Rosenblum and Kuttner 2006, p. 141) From my point of view, indeed, no physical force connects these microparticles; the wave corresponds to the two microparticles and this correspondence represents the instantaneous influence! As we will recall, correspondence is not a real phenomenon, i.e., it does not have any ontological status.

19 „EPR's argument assumed that the behavior of objects could be affected only by physical forces, and any object could otherwise be considered separate from the rest of the world. In particular, two objects could be separated so that the behavior of one could in no way affect the other." (Rosenblum and Kittner 2006, p. 141) Again, between these two microparticles there is, indeed, no physical force; there is only a correspondence between these two microparticles and the wave. However, the microparticles and the wave belong to EDWs.

20 As in cognitive neuroscience, in quantum mechanics it is quite common to talk about correlations. After experimenting with correlations between two photons (more specifically, their angle of polarization), physicists have reached the conclusion that their „twin photons indeed had identical polarization, identical stick angles. (In quantum theory, where polarization is observer created, the twin-photon correlation must be explained by a mysterious influence instantaneously exerted on a photon by the observation of its twin.)" (Rosenblum and Kuttner 2006, p. 145) Actually, physicists believe that in principle, „any two objects that have ever interacted are forever entangled. The behavior of one instantaneously influences the other. An entanglement exists even if the interaction is through each of the objects having interacted with a third object. In principle, our world has a universal connectedness." (Rosenblum and Kuttner 2006, p. 150) From my point of view, we cannot talk about the 'universal connectedness' of some or all objects (particles), but only about the correspondence between objects and waves. My experiment with the stick and the two corresponding electrons (microparticles) directly mirrors Bell's inequality. Bell's inequality is broken just because those two microparticles (electrons or photons, etc.) correspond to the wave. A real entanglement between two microparticles (a phenomenon that belongs to the EW of those microparticles) does not exist as a real physical phenomenon. Again, there is only a correspondence between the wave and the two microparticles, and, I must repeat, this correspondence is not a real physical state which belongs to the microparticles-EW or to the waves-EW! Within the EDWs perspective, it is much easier to understand Rosenblum and Kuttner's inquiry: Nonphysical influences: If there's a mind that's different from the physical brain, how does it communicate with the brain? This mystery reminds us of the connection between two quantum-entangled objects – of what Einstein called 'spooky actions' and Bohr called 'influences'. (p. 189) I will remind you that in cognitive neuroscience the great unsolved problem is the binding problem, which is quite similar to this entanglement problem from quantum mechanics. In both cases, there are phenomena that belong to EDWs. Moreover, the problem of mental causation (from the philosophy of mind) and spontaneous activations of certain neuronal patterns (that correspond to mental causation) are in the same situation: mental causation does not exist within the brain, while spontaneous activations of certain

detected polarization) show that Bell's inequality is violated: „When the experiments were done, Bell's inequality was violated. Bell's straw man was knocked down — as he expected it would be. Our world does not have both reality and separability — one, perhaps, but not both. And we immediately admit to not truly understanding what the world being unreal or having a universal connectedness would imply." (Rosenblum and Kuttner 2006, p. 143).[21] The consequence of these experiments is that the system of those two particles gains the property of non-locality.

According to the EDWs perspective, the two particles are in EW1 (the micro- or quantum-EW), which has the spatial coordinates of the whole cosmic space. In this EW, all micro-particles only interact with/observe other micro-particles. In EW1, the property of non-locality does not exist, so the two particles cannot have it. 'Non-locality' (which is in fact the continuity) is a property of a wave, which belongs to EW2. Again, EW2 also corresponds to the whole of cosmic space. The difference between two EDWs is given not by their spatio-temporal frameworks, but by their entities and the interactions among them. It is completely wrong to assign the property of non-locality to the relation between objects belonging to EW1. All we can say is that the wave corresponds to the system of particles.[22] Both Einstein et al. and those supporting the Copenhagen interpretation were mistaken, because they introduced epistemological properties (which only belong to EDWs)

neuronal patterns simply correspond to the will (mental state) of a human subject to raise her right hand, for instance.

21 „Bell's theorem has been called 'the most profound discovery in science in the last half of the twentieth century'. It rubbed physics' nose in the weirdness of quantum mechanics. As a result of Bell's theorem and the experiments it stimulated, a once 'purely philosophical' question has now been answered in the laboratory: There is a universal connectedness. Einstein's 'spooky interactions' do in fact exist. Any objects which have ever interacted continue to instantaneously influence each other. Events at the edge of the galaxy influence what happens at the edge of your garden." (Rosenblum and Kuttner 2006, p. 139) Again, within the EDWs perspective, this sort of 'universal connectedness' between two microparticles does not exist! That is, there is no direct, real connection between the two microparticles (by „direct, real connection", I refer to any physical phenomenon which belongs to the same micro-EW). The wave corresponds to those two microparticles and this correspondence mirrors this sort of connectedness! We have to remember that the wave does not exist for the microparticles (electrons, photons, etc.) and the microparticles do not exist for the wave. Moreover, the microparticles do not exist for the table (or for the stick, see below), the table (or the stick) does not exist for the microparticles.

22 „Mathematically, the state of these particles (which should perhaps be called wavicles) can't be described by six numbers (representing their position and velocity), but by a wavefunction, describing the extent to which they are in different places. This gives them properties both of traditional particles (they're either here or there) and of waves (they can be in several places at once in a so-called superposition)." (Tegmark 2014, p. 167) From my point of view, just as in the case of many other physicists working within the unicorn world, Tegmark is forced to invent dubious notions like wavicles to represents a mixture of phenomena which belong to EDWs! Moreover, there are no particular entities that have properties both of particles and waves, but there are microparticles and waves that belong to EDWs.

into the unicorn world. Thus, we can observe that the so-called hidden variables and non-locality or non-separability, which were introduced to 'save the phenomena' of the unicorn world, are empty concepts. Only the unicorn world and a one-to-many relationship have forced us to even consider von Neumann's idea of classical logic's revision (a pseudo-alternative) for understanding Bohr's complementarity. In their famous paper, Einstein, Podolsky and Rosen concluded that quantum mechanics is an incomplete description of reality. In my opinion, by mixing two EDWs, nothing can be complete.[23]

We can explain the 'non-locality' of the microparticles from an EDWs perspective. If we accept Einstein's idea concerning the rigidity objects, we also have to accept that the entities of each EW are rigid, i.e., any entity – except the I – exists only at its surface. Epistemologically different interactions represent the syntheses of the manifolds into epistemologically different entities. However, when we analyze macro-objects within the EDWs perspective, we do not ignore micro-forces, because they are part of an EDW. For instance, planets, waves and microparticles are rigid objects. Their interactions determine their own existences only at their surface. Quantum states are all rigid objects. A quantum wave and a quantum particle are rigid entities. The 'non-locality' of two electrons corresponds, in fact, to the rigidity of a wave, that actually refers to the indivisibility of the wave (which belongs to the EW2) and to the fact that the wave is not composed of (but corresponds to) various microparticles (which belong to the EW1). The movement of an electron corresponds to a wave. In EW1, a force that acts upon one electron does not act simultaneously on the other electron, because there is nothing that can surpass the speed of light in any EW. However, by acting on an electron, we also act on the corresponding waves, even if we do not observe this process. Only the rigidity (indivisibility) of the wave (which belongs to EW2) means that the signal acts simultaneously on both particles. However, this does not mean that EDWs form parallel worlds, 'many-worlds' or a multiverse (either the type proposed by

23 „The paper, now famous as EPR for 'Einstein, Podolsky, and Rosen', did not claim that quantum theory was wrong, just that it was incomplete. Quantum theory supposedly denied a physically real world, and thus required an observer created reality, only because it was not the whole story. EPR would show that you could, in fact, know a property of an object without observing it. That property, they argued, was therefore not observer created. The property was a physical reality that the incomplete quantum theory did not include." (Rosenblum and Kuttner 2006, p. 128) In other words, if „a physical property of an object can be known without its being observed, then that property could not have been created by observation. If it wasn't created by its observation, it must have existed as a physically reality before its observation. EPR needed to display only one such property to show quantum theory to be incomplete." (Rosenblum and Kuttner 2006, p. 135) From the EDWs perspective, an entity is observed/it interacts with any other entity from the same EW, so it is not necessary for human observation to ontologize the observed entity! These microparticles (and any other entity or phenomenon – except mental states, the I) exist before human beings observe them.

quantum mechanics, or the type suggested by the idea of hyperspace[24]). The idea of the hyperverse (i.e. of EDWs) is completely different from these notions in theoretical physics.

In order to better understand this problem, we can use an analogy, even if this analogy does not fit the physical nature of the objects involved in it. Imagine two people, X and Y, using two electronic microscopes at the same time, each of them observing an electron on either end of a long stick. Evidently, they both observe micro-particles, not the stick in itself. A third person, Z, using another stick, pushes one of the ends of the first stick. Because of this, X and Y observe new motions of their electrons at the same time, and inform each other immediately that this has happened. The process is repeated many times. Thus X and Y decide that these particular simultaneous motions of the electrons are not the result of coincidence. However, A believes that reality has some hidden variables that they are not aware of, while B thinks that the two electrons experience a kind of non-separability. In order to support his approach, B introduces the violation of Bell's inequality. Replacing the stick with a wave (keeping in mind that the stick is divisible, while the wave is indivisible), we have the classical case of the violation of Bell's inequality which, according to Penrose, proved the existence of quantum entanglements between physically separated particles. (Penrose 2004, p. 584)[25] The wave (which belongs to EW2) as an indivisible, rigid entity, corresponds to the two electrons (which belong to EW1 and have the properties of reality and separability, from Einstein's point of view). The forces which act on a part of the wave act simultaneously on the whole wave. The entanglement between two separated particles corresponds to the individuality, unity or rigidity of the wave.[26]

24 From the EDWs perspective, the idea of hyperspace (that presupposes the combination of Einstein's theory of relativity and quantum mechanics) is completely wrong simply because it represents a mixture of EDWs. (See Vacariu and Vacariu 2010, last chapter)

25 „If it is found that Bell's inequality is violated, one or both of the premises that lead to the inequality must be wrong. In other words, if Bell's inequality is violated in actual experiments, our world cannot possibly have both reality and also separability." (Rosenblum and Kuttner 2006, p. 142) From my point of view, the separability between the two photons is broken because of the rigidity (indivisibility) of their corresponding wave! In reality, the wave (which only corresponds to the two microparticles) in a sense represents the 'universal connectedness' or 'faster-than-light influences', but this connectedness is not a physical process which really exists between the two electrons. Again, the correspondence between the wave and the electrons has no ontological status, i.e., it does not really exist as an entity/process in any EW.

26 „Clauser's experiments ruled out, in physics terminology, 'local reality' or 'local hidden variables'. The experiments showed that the properties of objects in our world have an observation-created reality or that there exists a universal connectedness, or both." Rosenblum and Kuttner 2006, p. 148) Again and again, from the EDWs perspective, within the EW of microparticles, there is no real physical phenomenon taking place between the two photons. It is neither about the observation-created reality, nor about universal connectedness, but about the correspondence between those two

Einstein transformed an empirical law into a constitutive principle. From our point of view, the conditions of observation are elevated to the status of constitutive principles. However, they are constitutive because they help us to reveal the already existing EDWs. Within the hyperverse, the conditions of observation are replaced by conditions of constitutive interaction among entities belonging to EDWs. Maybe for philosophers like Carnap, Goodman, Quine and Putnam (and analytic philosophy in general), the question of what really exists makes no sense, it is a pseudo-question. Putnam assumes that the existence of the 'thing-in-itself' requires a view from an Archimedean point, namely a view corresponding to an ideal or impersonal knowledge. However, he claims that realism – which basically assumes the existence of one unique world – is not incompatible with conceptual relativism. (Putnam 1987, p. 17)

To support my approach, in this section I will continue to analyze certain notions and alternatives from the realm of physics. I want to offer a detailed demonstration of the fact that, by working within the unicorn world, physicists have been unable to explain certain spooky problems or mysteries in quantum mechanics in the last 100 years.[27] Through the EDWs perspective, we can clarify or reject these mysteries and (thought) experiments from quantum mechanics. However, I leave it to the specialists working in (the philosophy of) physics to develop, from an EDWs perspective, this new framework for quantum mechanics and for its relationship with Einstein's theory of relativity, as well as with other notions belonging to physics and cosmology.

(1) Young's experiment and Wheelers' delayed-choice experiment (1980)

We have to keep in mind that at all points of the experiment, as well as before, during and after we make our measurements, there is no unicorn world; the waves and the particles exist in EDWs within the hyperverse. Our observations depend on the tools of observation we use at a certain time.

If we take these elements into consideration, we can understand the interference pattern of the waves produced by electrons. Within the unicorn world, the interference we observed when we fired electrons at a screen was inexplicable. In this case, if we fire electrons (which belong to one EW) through the double-slit apparatus from Young's experiment, the screen displays the results of the interference of two waves (which belong to another EW). However, when one slit is closed, the screen displays only the results of the action of the electrons, but not of the wave. The very troubling question which therefore appeared in quantum

microparticles and the wave, which belong to EDWs! (I found it necessary to repeat this idea several times because it is very important for understanding the mysteries of quantum mechanics.)

27 See the article „100 Years of Quantum Mysteries" by Tegmark and Wheeler (2001).

mechanics in the last century was, 'Does this electron know whether the other slit is open or closed?' This is a pseudo-question. In fact, the wave passes through both slits, while the electron only passes through one slit.[28] Between the wave and the electron there is only a correspondence.

J. A. Wheeler, the famous theoretical physicist, embraces Hume's scepticism. Davies says of Wheeler that „He summed up his position with a typical Wheelerism: There is no law other than the law that there is no law". (Davies 2004, p. 6; 2006, p. 267) From one point of view, this law can be applied to the universe, the unicorn world, but not to each EW. As a reply to Hume's skepticism, we can say that for us the existence (over time) of epistemologically different entities in EDWs presupposes the existence of epistemologically different laws (interactions). If we deny the existence of all these laws, we deny our own existence. However, from another point of view, Wheeler is correct. He denies the existence of Platonic eternal laws of the universe. Physical laws are the result of the congealing of the universe after the Big Bang, and they are not eternally fixed (exactly the same) for all various forms of the world. (Davies 2006, p. 267) In this sense, one of Wheeler's main concepts is that of the mutability of laws.

Following Wheeler, Davies emphasizes the role of the experimenter (observer) in determining the nature of quantum reality in Young's experiment. Davies asks „When, exactly, did nature decide to opt for wave or particle?" (Davies 2006) Nature does not decide on any option simply because we observe it – more than that, nature, i.e., the unicorn world, does not exist. Although it is valid for the majority of physicists, Davies' inquiry is possible only within the unicorn world. Someone can talk about 'nature making decisions' only when nature is the unicorn world.

These issues have led to the creation of Wheeler's delayed-choice experiment, which is based on the idea that the past depends on the future. (Greene 2004, p. 186) In the split-beam experiment, a new photon detector is inserted immediately after the beam splitter. (p. 187) When the new detector is switched off, the photons produce interference patterns on a photographic screen. When the new detector is switched on, it indicates which path each photon travels. „Such 'which-path' information, as it's called, compels the photon to act like a particle, so the wavelike interference pattern is no longer generated." (pp. 187-8) If the distance between the beam splitter and the new detector is much larger, „the new weirdness comes from the fact that the which-path measurement takes place long after the photon had to decide at the beam splitter whether to act as a wave and travel both paths or to act as a particle and travel only one". (Greene 2004, p. 188) The anomaly seems to be

28 Evidently, the notion of slit is used here only for a pragmatic reason. In reality (i.e., the EDWs), an electron does not pass through a slit, but through an amalgam of microparticles which corresponds to that slit.

that the which-path measurement influences the past, i.e., the status of whatever entity passed through the beam splitter.

Again, within the unicorn world, we can find many anomalies. It is quite natural to consider that the wave and the particle cannot both be at the same place at the same time. In fact, the photon does not decide its situation before passing through the slit at all. Depending on our conditions of measurement, we can observe either the wave or the particle, which exist in EDWs before our observations take place.

We have to avoid confusing EDWs with the parallel universes theory. To explain the split-beam experiment, we do not need any shadow particles belonging to parallel universes. We can observe that Deutsch (and other physicists who follow Everett) are working within the unicorn world, even if they create many worlds or parallel universes. For Deutsch, these parallel universes exist at the same time in the unicorn world. As we saw above, we can easily explain this experiment by considering that the wave and the particle belong to EDWs, not to parallel universes.

(2) Heisenberg's uncertainty principle

The uncertainty principle, which tells us that we cannot measure the position and the velocity of a particle at the same time, is constructed within the unicorn world. It specifies that „[u]ncertainty is built into the wave structure of quantum mechanics and exists whether or not we carry out some clumsy measurement".[29] (Greene, p. 99)

I believe that this principle is based on the relationship between a wave and a particle, therefore there is a mixture between two EDWs. Measuring the location of an electron depends on the magnitude of its wave function. For instance, if a wave has a uniform succession of peaks and troughs, then the particle has a definite velocity. Nevertheless, its position is completely undetermined: the particle is equally probable to be anywhere within the wave. From an EDWs perspective, this is a mixture of two EDWs, that of the particle, and that of the wave. This is why we need to use probability calculus to discuss the correspondence between the wave and the particle.

From this point of view, the EDWs perspective is quite close to Bohm's theory, which follows De Broglie's earlier pilot wave interpretation. According to Putnam, this approach is the classical example for the hidden variable theory. (Putnam 2005, p. 622) It claims that particles have definite positions and momenta at all times. Their trajectories are continuous, and determined by a 'velocity field',

29 Embracing a deterministic view of the universe, Einstein claims that „I can't believe that God plays dice". (Tegmark and Wheeler 2001, p. 71)

while their initial positions and momenta are distributed randomly. We can identify the positions of these particles only by using quantum mechanical probability. (Putnam 2005, p. 622)

(3) Schrödinger's cat, decoherence and the 'many worlds' or multiverse approach

Bohr believed that the laws of the micro-cosmos and those of the macro-cosmos are different due to the different sizes of their entities.[30] (Greene 2004, p. 2003) In this context, Greene's question regarding the delimitation between the micro-cosmos and the macro-cosmos is 'Where exactly is this border?' If the researcher considers that both micro- and macro-particles are within the same world, the question cannot be answered. However, decoherence is the „bridge between the quantum physics of the small and the classical physics of the not-so-small by suppressing interference – that is, by diminishing sharply the core difference between quantum and classical probabilities". (Greene 2004, p. 209) The initiator of decoherence is Zeh (1970), whose ideas are forwarded by Joos, Zurek and others. Prior to our observation, there is a superposition of various states for a particle (let us say, the spin of a particle is simultaneously up and down). So there is a quantum uncertainty regarding the spin of that particle.

Tegmark and Wheeler explain how „the quantum gets classical". (Tegmark and Wheeler 2001, p. 73) For instance, quantum uncertainty is given by the superposition of the position of two states of a particle (up and down) and their corresponding wave. Schrödinger's equation predicts this coherent superposition, which can be mathematically illustrated through a density matrix. The wave function of the particle corresponds to a density matrix which has four peaks, two of which indicate a 50% probability of the particle to be either up or down, while

30 While Bohr's view is „that quantum mechanics and classical physics are complementary aspects of nature" (Dyson 2004, p. 76), I believe that quantum mechanics and classical physics are descriptions of the EDWs simply because nature does not exist! One EW does not exist for any other EW. „Copenhagen invokes the complementarity principle to confront a spooky aspect of observation: the instantaneous collapse of an object's wavefunction everywhere by an observation anywhere." (Rosemblum and Kuttner 2006, p. 107) „Niels Bohr realized that he had to confront the spooky connection of knowledge with physical phenomena in order to allow physicists to just get on with doing physics without becoming involved with philosophy. He arbitrarily asserted his principle of complementarity: The two aspects of a microscopic object, its particle aspect and its wave aspect, are complementary, and a complete description requires both contradictory aspects, but we must consider only one aspect at a time. We avoid the seeming contradiction by considering the microscopic system, the atom, not to exist in and of itself. We must always include in our discussion — implicitly at least — the different macroscopic experimental apparatuses used to display each of the two complementary aspects." (idem, p. 108) Not only do microscopic entities and macroscopic entities belong to EDWs, but waves and microparticles similarly belong to EDWs. However, for my EDWs perspective, „Bohr's complementarity" is an inaccurate notion since one set of entities that belongs to an EW does not exist for any other set of entities that belong to an EDW!

the other two indicate the interference of these two states. In this situation, „[t]he quantum state is still coherent“. (Tegmark and Wheeler 2001, p. 73)

According to Tegmark and Wheeler, quantum uncertainty is different from the uncertainty of classical probability, for instance the probability distribution of a coin toss. The density matrix of a coin toss only has the first two peaks, which represent the fact that the coin is either tails or heads, but we do not know which of the two it is, because we have not looked at it yet. There are no peaks for the interference process. The tiniest interaction with the environment transforms the coherent density matrix into the classical density matrix with only two peaks, which represent either tails or heads.

The interference pattern of those two states (up and down), or the coherent state, accomplishes decoherence. „The Schrödinger equation controls the entire process.“ (p. 73)[31] The standard interpretation is that the measurement process is an interaction between the observer and the observed particle. At this moment, the person cannot perceive this superposition because the interference pattern accomplishes decoherence.

The objects we encounter in our daily life are not isolated, interacting instead with other entities. For example, the book that I am currently reading is struck by photons and air molecules. Those microparticles disturb the coherence of the macro-object's wave function, thus rendering interference effects impossible. (Greene 2004, p. 210) „Once environmental decoherence blurs a wave function, the exotic nature of quantum probabilities melts into the more familiar probabilities of day-to-day.“ (p. 210) Because of decoherence, Schrödinger's cat cannot be both dead and live. However, Greene and other physicists are not content with this alternative, asking instead „how one outcome wins and where the many other possibilities go when that actually happens“. (Greene 2004, p. 212) Since the debate between Newton and Leibniz, the question „Which really exists, the particle or the wave?“ has not received a decisive answer, which is quite normal due to the framework of the unicorn world.

31 „(...) Compton showed that photons bounced off electrons as if they were each tiny billiard balls. On the other hand, to display interference, each and every photon or electron had to be a widely spread-out thing. Each photon, for example, had to go through both slits in a barrier. How can an object be both compact and spread out? Well, a wave can be either compact or spread out. (But, of course, it cannot be both at the same time.)“ (in Rosenblum and Kuttner 2006, p. 73) This reasoning can be valid only in the unicorn-world! A wave can be compact (i.e., a microparticle) and spread out (i.e., a wave) in EDWs. „If we are still going to put up with these damn quantum jumps, I am sorry that I ever had anything to do with quantum theory.“ (Erwin Schrodinger in Rosenblum and Kuttner 2006, p. 69) Essentially, from my point of view, these „damn quantum jumps“ (that is, the Bohr's jumps of microparticles from one level to another) correspond to the different peaks of the wave! Therefore, only with the EDWs perspective, we can explain these miraculous phenomena! For instance, an electron appears from and disappears in nothing, but it corresponds to the wave (that belongs to an EDW).

From an EDWs perspective, the superposition of the various states of a particle before measurement is a mistake which was created by extending the superposition of the wave and the particle. Putnam reminds us that Schrödinger's equation shows us a state given by the vector sum or superposition of a vector which represents both states of a particle (in my example, both up and down), which is mathematically expressed through the abstract 'Hilbert space'.

> And Problem One is what are we to make of a state which is a superposition of two states like this, two states in which a macroobservable has different values? [...] If we never observe such a state, why don't we? All interpretations of quantum mechanics are required to give an answer to that question. (Putnam 2005, p. 620)

Because the wave and the particle belong to EDWs, they have no superposition. Consequently, there is no superposition of the various states of any particle. Working within the unicorn world in the 1920's, some physicists have created the idea of the unobservable superposition of two states of a particle, which exists before the observation.

Putnam relates the above Problem One with Problem Two, 'the problem of Einstein's bed', the existence of the superposition of states in which macro-observables have different values. Putnam refers to this in this way because Einstein once described the situation in the following way: „Look, I don't believe that when I am not in my bedroom my bed spreads out all over the room, and whenever I open the door and come in it jumps into the corner". (Einstein in Putnam, p. 624) This means that Einstein rejected Von Neumann's collapse hypothesis.

Another hypothesis which supports the EDWs perspective is Dyson's, who denies the existence of gravitons in one of his articles. (Dyson 2004, pp. 88–9) The majority of physicists accept that the gravitational field must be a quantum field with gravitons associated to it. Dyson remarks that there are no arguments (either empirical or theoretical – or even thought-experiments) to support this idea. The tools using for detecting gravity can observe only classical gravitational waves produced by massive entities. If we lack even thought experiments which support quantum gravity, then the gravitational field is a 'pure classical field', and gravitons do not exist.

This hypothesis supports my approach. Gravity is produced by massive objects and if we think that a planet is composed of microparticles, we come to believe in microgravity, or in the existence of gravitons which produce gravity. However, microgravity and gravitons are empty notions. Gravity exists in a single EW, the world of macro-objects: if there were no macro-objects, space would have no curvature. It is impossible to introduce curvature at the quantum level. Even if such an idea is thinkable, it remains impossible to use – I will remind you of Hanna's comments on Kant's paragraphs (A239/B298-9 and A248/B305): „[...] empty concepts cannot be meaningfully applied by us either to noumenal objects or to objects of our sensory intuition, and in that sense they are impossible – that is, impossible to use". Hanna 2001, pp. 90–1)

Again, I must draw your attention to the fact that while there can be confusions between EDWs and the ideas of the 'many-worlds', the multiverse, or parallel universes from the field of quantum mechanics, these are two completely different approaches: the many-worlds or parallel universes are created within the unicorn world. Parallel universes are ontologically different universes, EDWs are epistemologically different worlds. The 'many-worlds' approach or the multiverse or parallel approach (created by Everett (1957), and forwarded by Zeh, Zurek, Deutsch (see point (1) above) and Tegmark) seems to be the closest alternative to the EDWs perspective.

The 'many-worlds' interpretation was created by Everett as an alternative to the collapse of the wave function into a particle during measurement (the Copenhagen interpretation).[32] According to Tegmark and Wheeler (2001), Schrödinger's equation predicts that the person seeing a particle will enter a superposition of two possible states. (p. 72) There are two parts of the total wave function (which includes the person and the particle) which work completely independently in two parallel worlds. Again, the many-worlds interpretation and the EDWs perspective are completely different approaches.

The idea of the superposition of two waves and that of the pseudo-superposition of the wave and the particle led physicists to the idea of the superposition of the various states of a particle before measurement. From an EDWs perspective, because the wave and the particle belong to EDWs, the particle does not have a superposition of various states. Thus, there is a totally different relationship between the parallel universes ('many-worlds' or multiverse) in the first approach and the different worlds of the EDWs approach. The parallel universes are said to ontologically exist in the unicorn world simultaneously, while EDWs epistemologically exist in the hyperverse. The number of parallel worlds can be huge[33], while the number of EDWs is limited, and it is given by the epistemologically different interactions between entities and by correspondences. Everett tried to solve the problem of superposition as a reply to the Copenhagen interpretation concerning the fact that the „wave function collapsed into some definite classical outcome whenever an observation was made, with probabilities given by the wave function". (Tegmark 2004, p. 473) For Everett, this „controversial collapse postulate was unnecessary". (Tegmark, p. 473) In fact,

32 However, from my point of view, Everett was right when he claimed that the „wavefunction never collapses. Ever." (Tegmark 2014, p. 171) Everett introduced the 'many worlds' interpretation simply because he did not notice the existence of EDWs which directly rejects the universe/unicorn world.

33 „I repeat, on the Many Worlds interpretation, there will be 2^{30} Einstein-'histories – parallel worlds'; science fiction is literally right!" (Putnam 2005, p. 630)

quantum theory alone predicted that one classical real scene would split into the superposition of many. Tegmark's remark that Everett could not solve two essential questions is interesting for the EDWs perspective:

1) Why do we not perceive macrosuperposition?
2) „What physical mechanism picks out approximately classical states (with each object in only one place, etc.) as special in the bewilderingly large Hilbert space?" (Tegmark 2004, p. 474)

Decoherence answers both questions. But decoherence is a pseudo-notion within the unicorn world. (See the last chapter of Vacariu 2008) The cat is not both dead and alive before our observation. This decoherence was only necessary to scientists because of the unicorn world. They consider that superpositions only exist in isolated systems. When these systems encounter other entities (such as a photon or molecules), there is a split between the parallel universes of those superpositions. Surprisingly, Tegmark wrote that „Decoherence is now quite uncontroversial and has been experimentally measured in a wide range of circumstances". (p. 474) Is he correct?

The general view concerning the articles which I have analyzed here is that physicists accept various, sometimes contradictory, alternatives (which can contain odd notions) in order to explain weird phenomena. At the end of their article (2001), Tegmark and Wheeler introduced the results of an informal pol at a conference on quantum computation at the Isaac Newton Institute (Cambridge, July 1999). Out of 90 physicists, 8 accepted the wave-function collapse, 30 preferred 'many-worlds or consistent histories (with no collapse)' and 50 accepted 'none of the above or undecided'. „Rampant linguistic confusion may contribute to that large number. It is not uncommon for two physicists who say that they subscribe to the Copenhagen interpretation, for example, to find themselves disagreeing about what they mean." (p. 75)[34]

Tegmark and Wheeler mention that quantum theory 'is probably just a piece in a larger puzzle'. Theories in physics can be organized in a family tree. General relativity and the quantum field theory can be found at the very top of the tree. However, „[p]hysicists know something is missing at the top of the tree, because we

34 Along the same lines, see Putnam's example. (Putnam 2005, p. 619) Putnam tried to convince his friend, a „world famous physicist", that there is a problem in quantum mechanics. Before several talks with Putnam, the physicist accepted the Copenhagen interpretation. Afterward, the physicist accepted the problem. Fourteen years later, at a conference he said: „There is no Copenhagen interpretation of quantum mechanics. Bohr brainwashed a generation of physicists." (p. 619) From my point of view, Bohr is no more guilty than other physicists. The fact that everybody has accepted the paradigm of the unicorn world is the biggest mistake which has brainwashed scientists and philosophers for such a long time!

lack a consistent theory that includes both gravity and quantum mechanics, yet the universe contains both phenomena". Therefore, the „ultimate goal of physics" is to find the „theory of everything" that „would have to contain no concepts at all". (Tegmark and Wheeler 2001, p. 75)

As we saw above, the theory of everything only has meaning within the unicorn world. To summarize my analysis from the EDWs perspective, I believe that the persistence of this peculiar picture of quantum mechanics for 100 years is due to the extension, within the unicorn world, of the correct idea of a waves' superposition to the pseudo-superpositions (1) of waves and particles and (2) of the several states of a particle.[35] Scientists and philosophers have obviously been forced to create such weird notions due to working within the paradigm of the unicorn world.[36]

I will end this section with Bell's view on quantum mechanics: during one conference

> with wit, and in his Irish voice, he firmly emphasized the depth of the unsolved quantum enigma. In big, bold letters on the blackboard he introduced his famous abbreviation, FAPP, „for all practical purposes", and warned against falling into the FAPPTRAP: accepting a merely FAPP solution. (Rosenblum and Kuttner 2006, p. 140)

Bell was right: quantum mechanics has been a very good theory only for all practical purposes. The problem is that reality is not the unicorn world in which this theory was constructed (indeed, reality has nothing to do with 'all our practical purposes'), but the EDWs. In order to avoid any FAPPTRAP, we need to exchange our old and false paradigm of thinking (the unicorn world in which various pseudo-interpretations of the problems of quantum mechanics have been elaborated) for a

35 „The relativity pioneer Roger Penrose quipped: 'There are probably more different attitudes to quantum mechanics than there are quantum physicists. This is not inconsistent because certain quantum physicists hold different views at the same time.'" (Tegmark 2014, p. 162) „The quantum enigma has challenged physicists for eight decades. Is it possible that crucial clues lie outside the expertise of physicists? Remarkably, the enigma can be presented essentially full-blown to non-scientists. Might someone unencumbered by years of training in the use of quantum theory have a new insight? After all, it was a child who pointed out that the emperor wore no clothes." (Rosenblum and Kuttner 2006, p. 13) Indeed, only a philosopher could have solved the 'quantum enigma'! (See the second motto of my book)

36 „Quantum mechanics tells us strange things about our world, things that we do not fully comprehend. This strangeness has implications beyond what we generally consider physics. We might therefore be tolerant when nonphysicists incorporate quantum ideas into their own thinking — even if they do so with less than complete understanding, or even a bit incorrectly." (Rosenblum and Kuttner 2006, p. 152) Indeed, quantum mechanics tells us „strange things about our world" simply because the world does not exist! Physicists cannot afford to be merely tolerant of the EDWs perspective (an approach elaborated by a philosopher). They have to accept and incorporate their empirical and theoretical works within this new framework of thinking, the EDWs perspective, in order to get much better results than in the past.

better framework, the EDWs perspective. Reality is the EDWs and separability of any two objects from the same EW has to be related with the correspondence of other phenomena from an EDW.[37] Contrary to Bohr's idea (one motto of this book, 'Everything we call real is made of things that cannot be regarded as real'), 'everything we call real' is made of things (epistemologically different entities) that are real and belong to EDWs, but the world/'Universe' is not real, i.e., the world does not exist. Accepting the existence of EDWs, physicists (and scientists in general) will surpass many great pseudo-problems that dominated their entire careers.

8.5. The relationship between Einstein's theory of relativity and quantum mechanics

„The more deeply you think of relativity, the less strange it seems. The more deeply you think of quantum mechanics, the more strange it seems." (Rosenblum and Kuttner 2006, p. 192) From my point of view, gravity and quantum forces belong to EDWs, so the attempt to unify these two theories is a false path. It is probable that a long time ago, (immediately after the Big Bang) at least some of these forces did not exist. Perhaps there were other EDWs than the current EDWs, in which the forces and entities were epistemologically different from the forces and entities we are aware of today. What is certain is that immediately after the Big Bang, there was no gravity in any EW which existed then. The appearance of large macrocosmic objects such as planets and stars meant the appearance of the macro-EW, an epistemologically different world from the micro-EW.

In the last years of his life, Einstein was almost entirely forgotten by the many physicists who were working in quantum mechanics, which was then a new framework in physics. The relationship between Einstein's theory of relativity and quantum mechanics (the theory of unification) has been a great problem in physics for many decades. Einstein's description of reality through the special theory of

37 „Along with Bell, we suspect that something beyond ordinary physics awaits discovery. Not all physicists would agree. Many would like to dismiss the enigma, our 'skeleton in the closet', as merely a psychological problem, claiming that we just have to get used to the quantum strangeness. However, the existence of an enigma is not a physics question. It's metaphysics in the original sense of that word." (Rosenblum and Kuttner 2006, pp. 87-88) Indeed, it was necessary for a philosopher to reject the existence of the world in order to solve the quantum enigma! Concerning quantum mechanics, John Wheeler said that, „Somewhere something incredible is waiting to happen." (in Rosenblum and Kuttner 2006, p. 124) This 'something incredible' is my discovery of EDWs which directly rejects the wrong notion of the world/Universe which dominated human thinking for millenniums. By replacing the unicorn world with EDWs, we solve the great 'quantum enigma' that dominated physics for almost a century!

relativity can be more easily understood and accepted within my EDWs perspective than within the unicorn world.

As I showed in my books (2008 and 2010) this relationship mirrors the relationship between micro-objects and macro-objects, that is, from my point of view, between the micro-EW and the macro-EW. One EW does not exist for any other EDW, so the relationship between the micro-EW and the macro-EW (or between quantum mechanics and Einstein's theory of relativity) is meaningless. As I have demonstrated in Vacariu (2008) and Vacariu and Vacariu (2010), quantum mechanics created a mixture of EDWs (the microparticle-EW and the wave-EW), and this is the main mistake of this paradigm.

Gravity exists neither in the micro-EW (gravitons do not exist), nor in macro-EW (Einstein was right, gravity is simply curved space). Moreover, from my point of view, macro-objects do not exist at all in the micro-EW (a microparticle does not interact with a macroparticle), therefore gravity does not exist.[38]

In this sense, I will comment on very recent news: by unifying Einstein's theory of relativity and quantum mechanics, Laura Mersini-Houghton (a professor at UNC-Chapel Hill) mathematically proves that black holes cannot exist. „The work not only forces scientists to reimagine the fabric of space-time, but also rethink the origins of the universe." (Benios, September 23, 2014) Physicists have believed that black holes collapse under their own gravity to a single point in space (i.e., the singularity). However, Einstein's theory of gravity predicts the existence of black holes, while quantum mechanics predicts that no information can disappear from the universe. (idem) „Many physicists and astronomers believe that our universe originated from a singularity that began expanding with the Big Bang. However, if singularities do not exist, then physicists have to rethink their ideas of the Big Bang and whether it ever happened." (Benios 23th September, 2014) In my books (2008, 2010, 2014), within the EDWs perspective, I showed that combining Einstein's theory of relativity with quantum mechanics is totally wrong. Therefore, either through combining these two theories (Mersini-Houghton's work) or through completely rejecting the combinations of these two theories (the EDWs perspective), black holes do not exist.

In Vacariu and Vacariu (2010, last chapter), we investigated some wrong notions created by physicists through the mixture of phenomena that belong to EDWs. On pages 239-295, within the framework imposed by the physicist Smolin, we analyzed black holes and we showed that, because we consider infinities (the infinite) to be merely tools created by human beings, there are great problems with the existence of black holes. „Now we can understand much easier the above speculation for the elimination of singularities (i.e., the infinites): the analysis of

38 For more about gravity, Newton vs. Einstein, quantum mechanics, and quantum gravity, see Vacariu and Vacariu (2010), Chapter 7.

matter using the quantum theory means that this matter belongs to the micro-EW, while the fact that the geometry of space and time is studied from a classical viewpoint means that these characteristics refer to the macro-EW. In fact, working in the unicorn-world and mixing two EDWs, the scientists have created a pseudo-problem." „For explaining the black holes, we need the general theory of relativity, the gravitational collapse produces the spacetime singularities. However, these singularities cannot be explained by the general theory of relativity, the principle of equivalence being not valid. (Hendrich 2006) At such dimensions, maybe we can use quantum mechanics. Nevertheless, using both theories great contradictions cannot be avoided." (pp. 304-5) „Greene underlies the philosophical reason for combining the theory of general relativity and quantum mechanics: „If we ever hope to understand the origin of the universe – one of the deepest questions in all of science – the conflict between general relativity and quantum mechanics must be resolved. We must settle the differences between the laws of the large and the laws of the small and merge them into a single harmonious theory." (Greene 2004, p. 338, his emphasize) Within the unicorn-world, we really need indeed to solve this infamous conflict. Such Leibnizian harmony is possible only within the unicorn-world." (p. 305) „We saw at point (d) that some paradoxes regarding the black holes have appeared just because of the mixture of two EDWs." (Vacariu and Vacariu 2010, p. 297) „The dark matter and the dark energy[39], the infinities in physical phenomena or even the black holes are other Ptolemaic epicycles created by the human scientific imagination." (Vacariu 2014, p. 318)

Let me make two analogies: (1) Between black holes and gravity: from the EDWs perspective, the planets (which belong to the macro-EW) do not exist for microparticles (which belong to the micro-EW). Therefore, the curvature of the spatio-temporal framework (the gravity) caused by the planets does not exist for microparticles (as we indicated in Vacariu and Vacariu 2010, quantum gravity cannot exist). There is only something (micro-phenomena/processes) which corresponds to this macro-curvature (which belongs to the macro-EW). By analogy, if they exist, the so-called black holes (very strange things which belong to the macro-EW alone) only correspond to some phenomena/entities that belong to an EDW (which I will refer to as the macro-macro-EW). Moreover, such black holes cannot be singularities, since the infinite, as I have shown in one of my

39 The relationship between dark energy/matter and galaxies is quite similar to the relationship between microparticles and waves. It has to be a correspondence between the accelerating galaxies and some phenomena from an EDW than the macro-EW of galaxies, planets, etc. The black holes from our macro-EW correspond to some phenomena that belong to an EDW. Therefore, black holes are not the gate toward another world. Some physicists (for instance Krauss in his book „A universe from nothing", 2012) consider that the universe appeared from nothing. From an EDWs perspective, the universe appeared from hypernothing. (See my future work on this topic and other related topics such as the Big Bang, dark matter and dark energy, inflation, etc.)

presentations (see my webpage) does not exist, it is just a creation of the human mind. (2) Between black holes and mental causation (the mind moves the body's arm, for instance): it is believed that there are downward processes, i.e., the mind acts on the body, in cases such as moving the arm of that same body. From my point of view, mental causation and downwards processes such as this one do not exist. The mind does not act on the body's arm, since the mind does not exist for the brain (and the entire body) and vice-versa. The mind acts on a virtual arm that corresponds to the real arm. Only the brain acts on the real arm of the body placed within an environment. (In this sense, see Vacariu (2008), in which I investigate Ramachandran's famous case of phantom limb, where an individual feels a pain in his amputated arm.)

It was important to discover the existence of EDWs and to show that the theory of relativity perfectly explains truly existent phenomena in the macro-EW (which is neither an approximation, nor an appearance of reality), while quantum mechanics represents a pseudo-theory (before the discovery of EDWs, microparticles were considered to be the real components of the universe) which mixes phenomena belonging to EDWs (the wave-EW and the particle-EW).

Therefore, I attest Einstein's absolute revenge: his theory explains some real phenomena/processes which belong to EDWs that really exist. Quantum mechanics does not explain reality, since the unicorn world does not exist.[40] Indeed, physicists need to change their framework of thinking so as to avoid empty results in the future.[41]

I believe that now is the time to change the old paradigm of the unicorn-world with a new one, the EDWs paradigm, not only in philosophy, but also in science. Woit quotes the famous mathematician Michael Atiyah („Atiyah is one of the greatest mathematicians of the second half of the twentieth century" – Woit 2006, p. 265) who gave a talk entitled „The interaction between geometry and

40 „In the EPR argument, Bell, Clauser, and Aspect showed Bohr to be right and Einstein wrong. But Einstein was right that there was something to be troubled about. It was Einstein who brought quantum theory's full weirdness up front. It was his objections that stimulated Bell's work and that continue to resonate in today's attempts to come to terms with the strange worldview quantum mechanics forces on us. Bell believes that in 'his arguments with Bohr, Einstein was wrong in all the details. Bohr understood the actual manipulation of quantum mechanics much better than Einstein. But still, in his philosophy of physics and his idea of what it is all about and what we are doing and should do, Einstein seems to be absolutely admirable... [T]here is no doubt that he is, for me, the model of how one should think about physics.'" (Rosenblum and Kuttner 2006, p. 151) With the EDWs perspective, I showed that Einstein was in a better position than Bohr et. Company!

41 The title of a section of Chapter 13 from Rosenblum and Kuttner (2006) is „Is It Einstein for Whom the Bell Tolls?" (p. 151) From my point of view, Einstein was right regarding the weirdness of quantum mechanics. I am sure Einstein would be delighted to read about the EDWs perspective and its application to the 'quantum enigma'!

physics" during a conference on „The unity of mathematics" at Harvard University in September 2003.

> If we end up with a coherent and consistent unified theory of the universe, involving extremely complicated mathematics, do we believe that this represents reality? Do we believe that the laws of nature are laid down using the elaborate algebraic machinery that is now emerging in string theory? Or is it possible that nature's laws are much deeper, simple yet subtle, and that the mathematical description we use is simply the best we can do with the tools we have? In other words, perhaps we have not yet found the right language or framework to see the ultimate simplicity of nature. (Atiyah 2003 in Woit 2006, p. 265)

Indeed, in physics, it is necessary to find a new global framework. I believe that this needs to be the EDWs perspective. Moreover, this type of new framework of science and philosophy entails a new framework of thinking in general, because the world has dominated our vision of the realities surrounding us since our species first began to think. This type of change is the most difficult thing to achieve, especially for those who are already too deeply trapped in the old paradigm. My perspective is neither, as Atiyah mentioned above, about the simplicity of nature, nor about the multiplicity of the multiverse, but about EDWs. Nature is not simple, but subtle. It is not the unicorn-world, but the EDWs. My approach might be negative in rejecting so many things in philosophy and science, but it is also positive in proposing the requisite new framework, the EDWs perspective. My message is mainly for young scientists and philosophers who prefer not to waste their time doing research under the umbrella of the unicorn-world.

Conclusion

Contemporary philosophy does not exist for the physicists (and the scientists in general) of the last century. Could great contemporary physicists, people working in cognitive (neuro)science, biology and other domains accept the EDWs perspective proposed by an unknown philosopher? As we saw in the Introduction, scientists believe that „philosophy today is dead" (Hawking, 2010). As far as I am concerned, they are absolutely right, since contemporary philosophers have nothing to do with the elaboration of the necessary Weltanschauung for particular sciences, all of which are surrounded by great problems. Scientists might be confronted with these problems today, but they are not expecting any help from philosophers, since there has been an increasingly large gap between science and philosophy which appeared at least 150 years ago. The gap is due to the exponential growth of the knowledge available in the particular sciences, knowledge which is necessary, but nearly impossible to obtain by a philosopher who is meant to elaborate a Weltanschauung of all these sciences.

Nevertheless, the main problems in particular sciences have been created by people working within the unicorn world. My desire is to furnish the new Weltanschauung, which will transform many such problems into pseudo-problems. Therefore, my message is mainly for scientists, not for actual philosophers. My EDWs perspective is, as you will discover at even the most cursory of glances, completely different from any kind of philosophy in the past 150 years.

I did not grow up under the umbrella of contemporary philosophy, and this was the necessary condition for me to elaborate the EDWs perspective as a new framework of thinking for people working in cognitive (neuro)science, as well as for physicists, biologists, and young philosophers. It is understandable that contemporary philosophers and a large number of scientists cannot accept my EDWs perspective, since all of their achievements during their entire careers would be entirely invalidated.[1] If scientists insist upon applying Occam's razor (since it is

1 Reminding again Max Born's opinion about quantum mechanics (see one of the motto of this book), we can conclude now, at the end of this book, that the great problems of particular sciences are indeed philosophical problems. Such problems require, in general, changing an old paradigm with a new one, a task that is the job of a philosopher-scientist (Descartes or Kant, for instance) or a scientist-philosopher (Newton, Einstein or Bohr, for instance). In this context, we can re-think a little Planck's opinion on changing paradigms (See one of motto of this book): If one wants to change a

easier for some of us to think of one world rather than many EDWs), then the futile process of fabricating very knotty Ptolemaic epicycles for pseudo-problems, such as the mind-body problem, the nature of the world, the relationship between microscopic and macroscopic entities, quantum mechanics, inflation, 'levels of reality' or the superstring theory, will continue.

The philosophy of the EDWs is a step forward towards overcoming the scorching pseudo-disputes of the last century. In order to avoid wasting time working on such pseudo-problems by creating marvelous Ptolemaic epicycles, people now have the opportunity to replace the unicorn world with EDWs. Obviously, the final decision belongs to each of them. Nevertheless, I would like to mention that it was no accident that I selected Hawking's words as the motto of this work. The unicorn world was the most general framework, which generated the largest umbrella of illusory knowledge. What kind of knowledge could we have obtained by working within a world that does not exist? Only local knowledge, which explains local phenomena belonging to EDWs.

Smolin wrote about the superstring theory: „If the new dimensions and symmetries do not exist, then we will count string theorists among science's greatest failures, like those who continued to work on Ptolemaic epicycles while Kepler and Galileo forged ahead". (Smolin 2006, p. xvii) Those who continue working within the unicorn world will create new Ptolemaic epicycles, those embracing the EDWs perspective will forge ahead. The EDWs perspective changes the largest Weltanschauung in the history of human thinking, trashing the greatest Ptolemaic epicycle, that of the world (the most tangible, but because of this reason, the most dangerous notion). Therefore, in Nietzsche' style I finally proclaim:

The world is dead.
Long live the EDWs.

paradigm, one has to wait until the famous (and usually old) persons working in the paradigm's domain retire or dye. We can only hope that, with today's dynamics of knowledge, changed as they are by the Internet, it will not take so long to obtain a paradigm shift. In any case, EDWs are not for today, the place of the remnants of the last century, but for tomorrow, the time of new contretemps!

Glossary

Axiomatic-hyperontological framework is a axiomatic framework that furnishes the ontological status of all various objects that exist in EDWs

Cognitive neuroscience is the particular science concerned with the relationship between the mind and the brain (between any mental state and neuronal patterns). This science has been created through the unification of certain types of information (of certain notions) in psychology and neuroscience.

Correspondences refer to the abstract relationships between the epistemologically different objects which belong to EDWs. These correspondences are abstract (they do not exist in reality), since no EW exists for any other EDW.

Dualism is a notion belonging to the French philosopher Descartes, and it refers to the fact that the mind and the body (which includes the brain) are two ontologically different substances. Descartes was unable to show how these two substances interact.

Epistemologically different interactions refer to the relationships between epistemologically different objects. A particular case of this type of relationships are physical forces. We can make an (incorrect) analogy between interactions and perceptions: a table exists for me because I see it in front of me; that table exists for the coffee mug which stands (interacts due to the gravity between the table and the coffee mug) on the table precisely because that mug perceives the table it is standing on (and the table perceives the mug in its turn).

Epistemologically different objects: An object exists because it interacts with other objects in the same EW. Namely, that object exists only for the objects it interacts with from the same EW, not for the epistemologically different objects which belong to EDWs.

Epistemological world (EW) is a set of objects/entities and their interactions. The objects/entities which interact belong to the same EW.

Epistemologically different worlds (EDWs) are sets formed of epistemologically different objects/processes and the interactions between them. No EW exists for any other EDW.

The Hyperverse is just a label for all EDWs, i.e., hyperverse does not represent something that exist as a whole since an EW does not exist for any other EW

The I/self/mind/life corresponds to an organism (body) or to a cell. The I is an EW without any spatial dimensions. The I is, on the one hand, made up of mental states and, on the other hand, an entity which possesses unity precisely because it has no spatial dimensions. This is why the I is an indeterminate individuality.

The ontological-epistemological threshold represents the threshold for observational conditions which must be passed in order for the human observer to go from observing an entity belonging to EW1 to observing another entity (or a mixture of entities) which belongs to EW2. The entity in EW1 corresponds to the mixture of entities (and their interactions) which belong to EW2.

The organizational threshold represents the threshold for observational conditions which must be passed in order for the human observer to be able to move from observing, for example, an entity, to observing a set of entities. For example, a forest represents a whole for a human observer who is far away from it. For the same observer, who is at a very small distance from the forest, the forest no longer exists, being replaced with the trees which form the forest. Both the forest and the trees are in the same EW, but the forest does not exist for the trees, while the trees do not exist for the forest.

Quantum mechanics is an area of physics concerned with the study of microparticles and of the forces between them.

The theory of relativity, proposed by Einstein, is concerned with the study of macroparticles (stones, planets, etc.) and of the (gravitational) force between them.

The unicorn world refers to the world or the universe (all entities and processes are placed within the same world) which does not exist, being only a human intellectual creation which has been the entirely wrong framework used since the dawn of human though, up until now.

The Weltanschauung represents 'a philosophical image of the world', or a philosophical system about everything that exists, regardless of whether we have knowledge concerning what exists or not.

Bibliography

Anderson, P. W. (1972), More is Different. American Association for the Advancement of Science, 177, pp. 393–396

Aydede, Murat (2004), The language of thought hypothesis. The Stanford Encyclopedia of Philosophy (Fall 2004 Edition), Edward N. Zalta (ed.), URL = http://plato.stanford.edu/archives/fall2004/entries/language-thought/

Baars, J. Benjamin and Franklin, S. (2007), An Architectural Model of Conscious and Unconscious Brain Functions: Global Workspace Theory and IDA. Neural Networks, 20, pp. 955–961

Baars, J. Bernard (1988), A Cognitive Theory of Consciousness, Cambridge, MA, Cambridge University Press

Baars, J. Bernard and Gage, M. Nicole (2010), Cognition, Brain and Consciousness – Introduction to Cognitive Neuroscience, Second edition, Elsevier Ltd.

Banich, T. Marie and Compton, J. Rebecca (2011), Cognitive Neuroscience, (Third edition), Wadsworth, Cengage Learning

Bassett, S. Danielle and Gazzaniga, S. Michael (2011), Understanding complexity in the human brain. Trends in Cognitive Sciences, 15/5, pp. 200–209

Bechtel, William (1998), Representations and cognitive explanations: Assessing the dynamicist's challenge in cognitive science. Cognitive Science, 22(3), pp. 295–318

Bechtel, William (2002), Decomposing the Mind-Brain: A Long-Term Pursuit. Brain and Mind, 3, pp. 229–242

Bechtel, William (2008), Mental Mechanisms, Philosophical Perspectives on Cognitive Neuroscience, Routledge Taylor & Francis Group, LLC.

Bechtel, William (2009), Explanation: Mechanism, Modularity and Situated Cognition. In P. Robbins and M. Aydede (eds.), Cambridge Handbook of Situated Cognition, Cambridge, Cambridge University Press

Bechtel, William (2013), The endogenously active brain: the need for an alternative cognitive architecture. Philosophia Scientiae, 17/2, pp. 3–30

Beer, Rodney (1995), Computational and Dynamical Languages for Autonomous Agents. In R. Port and T. van Gelder (eds.), Mind as Motion: Explorations in the Dynamics of Cognition, MIT Press, Cambridge, MA

Benios, Thania (2014), Rethinking the origins of the universe. At http://unc.edu/spotlight/rethinking-the-origins-of-the-universe/

Bickle, John (2007a), The Philosophy of Neuroscience. Stanford Encyclopedia of Philosophy

Bickle, John (2007b), Who Says You Can't Do a Molecular Biology of Consciousness?. In Maurice Schouten and Huib Looren de Jong (eds.), The Matter of the Mind Philosophical Essays on Psychology, Neuroscience, and Reduction, Blackwell Publishing.

Bickle, John (2008), Real Reductionism in Real Neuroscience: Metascience, Not Philosophy of Science (and Certainly Not Metaphysics.). In J. Hohwy and J. Kallestrup (eds.), Being Reduced, Oxford, Oxford University Press, pp. 34–51

Black, B. Ira (1991), Information in the Brain: A Molecular Perspective, MIT Press, Cambridge, MA

Bressler, Steven L. and Menon, Vinod (2010), Large-scale brain networks in cognition: emerging methods and principles. Trends in Cognitive Sciences, 14, pp. 277–290

Brooks, A. Rodney (1991), Intelligence without representation. Artificial Intelligence, 47, pp. 139–159

Carnap, Rudolf (1950), Empiricism, semantics, and ontology. Reprinted from review International du Philosophie, iv (1950), pp. 20–40 in Richard M. Rorty (ed.), The Linguist Turn, The University of Chicago Press, 1967.

Chafee, V. Matthew and Crowe, A. David (2013), Thinking in spatial terms: decoupling spatial representation from sensorimotor control in monkey posterior parietal areas 7a and LIP. Frontiers in Integrative Neuroscience, 6, art. 112

Chalmers, J. David (2003), Consciousness and its place in nature. In S. Stich and T. Wartfield (eds.), Blackwell Guide to the Philosophy of Mind, Blackwell.

Chalmers, J. David (2006), Strong and weak emergence. In: Clayton P and Davies P (eds.) The Re-emergence of Emergence, Oxford University Press

Churchland, S. Patricia (1996), Toward a neurobiology of the mind. In Patricia S. Churchland and Rudolfo Llinas (eds.), The Mind-Brain Continuum: Sensory Processes, MIT Press, Cambridge, MA

Churchland, M. Paul (1998), Conceptual similarity across sensory and neural diversity: the Fodor-Lepore Challenge answered. The Journal of Philosophy, 95/1, pp. 187–207

Churchland, M. Paul and Churchland, S. Patricia (1990), Intertheoretic Reduction: a Neuroscientist's Field Guide. The Neuroscience 2, pp. 249–56 reprinted in R. Warner and T. Szubka (eds.), The Mind-Body Problem, Blackwell Publishing Inc, 1993

Churchland, M. Paul and Churchland, S. Patricia (1997), Recent work on consciousness: philosophical, theoretical, and empirical. In Churchland M. Paul and Churchland S. Patricia: 1998, On the Contrary. Critical Essays, 1987–1997, A Bradford Book, The MIT Press

Churchland, M. Paul and Churchland, Patricia S. (1998), On the Contrary. Critical Essays, 1987–1997, A Bradford Book, The MIT Press

Clark, Andy (1993), Associative Engines Connectionism, Concepts, and Representational Change, A Bradford Book, The MIT Press, Cambridge, Massachusetts, London, England.

Clark, Andy (1997a), Being There: Putting Brain, Body and World Together Again, MIT Press, Cambridge, MA

Clark, Andy (1997b), From Text to Process – Connectionism's Contribution to the Future of Cognitive Science. In David Martel Johnson and Cristina E. Erling, The Future of Cognitive Revolution, Oxford University Press

Clark, Andy (2001), Mindware – An Introduction to the Philosophy of Cognitive Science, New York, Oxoford, Oxford University Press

Cooper, P. Richard and Shallice, Tim (2010), Cognitive Neuroscience: The Troubled Marriage of Cognitive Science and Neuroscience. Topics in Cognitive Science, 2, pp. 398–406

Crane, Tim (2001), The significance of emergence. In Barry Loewer and Grant Gillett (eds.), Physicalism and its Discontents, Cambridge University Press

Crick, Francis and Koch, Christof (1997), Towards a neurobiological theory of consciousness. In N. Block, O. Flanagan and G. Guzeldere (eds.), The Nature of Consciousness, Cambridge, MA: MIT Press, pp. 277–292

Çukur, Tolga; Nishimoto, Shinji; Huth, G. Alexander and Gallant, L. Jack (2013a), Attention during natural vision warps semantic representation across the human brain. Nature neuroscience, online publication.

Çukur, Tolga; Huth, G. Alexander; Nishimoto, Shinji and Gallant, L. Jack (2013b) Functional Subdomains within Human FFA. The Journal of Neuroscience, 33(42), pp. 16748–16766

Damasio, R. Anotonio (1988), Time-Locked Multiregional Retroactivation: A System Proposal for the Neural Substrates of Recall and Recognition. Cognition, no. 33, pp. 25–62.

Davies, Paul (2004), John Archibald Wheeler and the clash of ideas. In D. Barrow, John Davies, C. W. Paul, Harper, L. Charles: 2004, Science and Ultimate Reality: Quantum Theory, Cosmology and Complexity, Cambridge University Press

Davies, Paul (2006), The Physics of Downward Causation. In Clayton, Philip and Davies, Paul (eds.) The Re-Emergence of Emergence: the Emergentist Hypothesis from Science to Religion, Oxford University Press: Oxford

Derrfuss, J. and Mar, R. A. (2009) Lost in localization: the need for a universal coordinate database. Neuroimage, 15/48(1), pp. 1–7

Dyson, J. Freeman (2004), Thought-experiments in honour of John Archibald Wheeler. In D. John Barrow, C. W. Paul Davies, L. Charles Harper: 2004, Science and Ultimate Reality: Quantum Theory, Cosmology and Complexity, Cambridge University Press

D'Esposito, Mark (2010), Why Methods Matter in the Study of the Biological Basis of the Mind: A Behavioral Neurologist's Perspective. In Patricia A. Reuter-Lorenz, Kathleen Baynes, George R. Mangun, and Elizabeth A. Phelps, The Cognitive Neuroscience of Mind - A Tribute to Michael S. Gazzaniga, A Bradford Book The MIT Press Cambridge, Massachusetts London, England

Dietrich, Eric and Markman, B. Arthur (2003), Discrete thoughts: Why cognition must use discrete representations. Mind and Language, 18/1, pp. 95–119

Diaz, Jose-Louis (2000), Mind-body unity, dual aspect, and the emergence of consciousness. Philosophical Psychology, 13/3, pp. 393–403

Dyson, K. Freeman (2004), Thought-Experiments in Honour of John Archibald Wheeler. In D. John Barrow, C. W. Paul Davies, L. Charles Harper, 2004, Science and Ultimate Reality: Quantum Theory, Cosmology and Complexity, Cambridge University Press.

Einstein, Albert (1982), How I created the theory of relativity. Physics Today, (August 1982) (translated by Yoshimasa A. Ono) (Einstein 1922)

Einstein, Albert (2011), Teoria relativitatii pe intelesul tuturor, (Theory of relativity for everybody), Humanitas

Fisher, W. Kurt and Bidell, R. T. (1998), Dynamic development of psychological structures in action and thought. In W. Damon (chief-ed.), Handbook of Child Psychology, Fifth edition, Vol. 1: Theoretical Models of Human Development, R. M. Lerner (vol. ed.), John Wiley & Sons, Inc.

Fodor, A. Jerry (1974), Special sciences or the disunity of science as a working hypothesis. Synthese 28, pp. 77–115

Fodor, A. Jerry & Pylyshyn, W. Zenon (1988), Connectionism and cognitive architecture. Cognition, 28, pp. 3–71

Fowler, F. Colin (1999), Descartes On the human Soul: Philosophy and the demands of Christian Doctrine, Kluwer Academic Publishers, Dordrecht, Boston, London

Friedman, Michael (1992), Kant and Exact Sciences, Cambridge, Mass., Harvard University Press

Friedman, Michael (2001), Dynamics of Reasoning, CSLI Publications, Standford, California.

Frith, Chris (2007), How the Brain Creates our Mental World, Blackwell Publishing.

Fodor, A. Jerry and Pylyshyn, W. Zenon (1988), Connectionism and Cognitive Architecture. Cognition, 28, pp. 3–71

Gabriel, Markus (2013) Warum es die Welt nicht gibt, Ullstein: Berlin

Gabriel, Markus (2013) Why the world does not exist.
https://www.youtube.com/watch?v=hzvesGB_TI0&feature=youtube_gdata_player

Gazzaniga, S. Michael (2010), Neuroscience and the correct level of explanation for understanding mind – An extraterrestrial roams through some neuroscience laboratories and concludes earthlings are not grasping how best to understand the mind–brain interface. Trends in Cognitive Sciences, 14, pp. 291–292

Globus, G. Gordon (1992), A noncomputational theory. Journal of Cognitive Neuroscience, 4

Globus, G. Gordon (1995), The Postmodern Brain, John Benjamins Greene, Brian: 1999, The Elegant Universe, Vintage Books

Greene, Brain (1999, 2003), The Elegant Universe: Superstrings, Hidden Dimensions, and the Quest for the Ultimate Theory

Greene, Brian (2004), The Fabric of Cosmos; Space, Time and the Texture of Reality, Vintage Books, New York

Greene, Brain (2011), Universul elegant, Humanitas Publishing Company (translation in Romanian of The Elegant Universe: Superstrings, Hidden Dimensions, and the Quest for the Ultimate Theory, 2003)

Gunzelmann, Glenn and Lyon, R. Don (2011), Representations and processes of human spatial competence. Topics in Cognitive Science,3, pp. 741–759

Hanna, Robert (2001), Kant and the Foundations of Analytic Philosophy, Clarendon Press, Oxford University Press

Haynes, John-Dylan (2009), Decoding visual consciousness from human brain signals. Trends in Cognitive Sciences, 13/5, pp. 194–202

Heil, John (2004), Philosophy of Mind, second edition, Routledge, NY and London

Hawking, Stephen and Mlodinow, Leonard (2010) The Grand Design. The Random House Publishing Group

Horst, Steven (2007), The computational theory of mind. The Stanford Encyclopedia of Philosophy, (Fall 2005 Edition), Edward N. Zalta (ed.),
URL = http://plato.stanford.edu/archives/fall2005/entries/computational-mind/

Huth, G. Alexander; Nishimoto, Shinji; Vu, T. An and Gallant, L. Jack (2012), A continuous semantic space describes the representation of thousands of object and action categories across the human brain. Neuron 76, pp. 1210–1224

Kaku, Michio (1994), A Scientific Odyssey Through Parallel Universes, Time Warps, and the 10th Dimension, Oxford University Press

Kant, Immanuel (1958), The Critique of Pure Reason, trans. N. K. Smith, New York, Modern Library

Karmiloff-Smith, Annette (1994), Précis of beyond modularity: A developmental perspective on cognitive science. Brain and Brain Sciences, 17, pp. 639–745

Kauffman, Stuart (1995), At Home in the Universe, New York, Oxford University Press

Kauffman, Stuart (2000), Investigations, Oxford University Press

Kauffman, Stuart (2008), Reinventing the Sacred: A New View of Science, Reason, and Religion. Basic Books

Kelso, J. A. Scott (1995), Dynamic Patterns, MIT Press/Bradford Books, Cambridge, MA and London, U.K.

Kim, Jaegwon (2005), Physicalism or Something Near Enough, Princeton University Press

Kim, Jaegwon (2006), Emergence: Core ideas and issues. Synthese, 152/3, pp. 547–559

Kosslyn, S. Michael (1997), Mental Imagery. In Michael S. Gazzaniga (ed.), Kosslyn, S. Michael; and Smith, E. Eduard, 2001, Higher Cognitive Functions – Introduction. In Michael S. Gazzaniga, (ed.), Cognitive Neuroscience, second edition, MIT Press

Kosslyn, S. Michael and Keonig, O. (1992), Wet Mind- the New Cognitive Neuroscience, The Free Press

Krauss, Lawrence (2012), A Universe from Nothing – Why There is Something than Nothing, FREE PRESS - A division of Simon & Schuster, Inc

Ionicioiu, Radu and Terno, R. Daniel (2011), Proposal for a quantum delayed-choice experiment. Physical Review Letters, 107, 230406

Laureys, Steven; Boly, Melanie and Tononi, Giulio (2009), Functional neuroimaging. In Laurey, Steven and Tononi, Giulio (2009), The Neurology of Consciousness: Cognitive Neuroscience and Neuropathology, Elsevier Ltd

LaRock, Eric (2010), Cognition and Consciousness: Kantian Affinities with Contemporary Vision Research. Kant-Studien 101. Jahrg., pp. 445–464

Lorenz, Konrad (1941), Kant's Doctrine of the A Priori in the Light of Contemporary Biology. In H. Plotkin (ed.) Learning, Development and Culture, Chichester, John Wiley and Sons, 1982

Lungarella, Max and Sporns, Olaf (2006), Mapping Information Flow in Sensorimotor Networks. Public Library of Science Computational Biology, 2/10, pp. 1301–12

Mandler, Jean (1998), Representation. In W. Damon (chief-ed.), Handbook of Child Psychology, Fifth edition, vol. 2: Cognition, Perception, and Language, Deanna Kuhn and Robert S. Siegler (vol. eds.), John Wiley, London

Markman, B. Arthur and Dietrich, Eric (2000), Extending the classical view of representation. Trends in Cognitive Sciences 4/1 2, pp. 470–475
Merzenich, M. Michael and Christofor, R. deCharms (1996), Neural representations, experience and change. In Rodolfo Llinas and Patricia S. Churchland (eds.), The Mind-Brain Continuum: Sensory Processes, MIT Press, Cambridge, MA
Moser, I. Edvard; Corbetta, Maurizio; Desimone, Robert; Frégnac, Yves; Fries, Pascal; Graybiel, M. Ann; Haynes, John-Dylan; Itti, Laurent; Melloni, Lucia; Monyer, Hannah; Singer, Wolf; von der Malsburg, Christoph and Wilson, A. Matthew (2010) Coordination in Brain Systems. In von der Malsburg, Christoph, Phillips A. William, and Singer, Wolf (2010)
Nagel, Thomas (1993), Consciousness and objective reality. In R. Warner and T. Szubka (eds.), The Mind-Body Problem, Blackwell Publishing Inc, 1993
Nishimoto, Shinji; Benjamini, Yuval; Vu, An T.; Yu, Bin; Naselaris, Thomas and Gallant, Jack L., (2011), Reconstructing Visual Experiences from Brain Activity Evoked by Natural Movies. Current Biology, 21, pp. 1641–1646
Northoff, Georg (2011b), Neuropsychoanalysis in Practice: Brain, Self, and Objects, Oxford University Press
Northoff, Georg (2014), Unlocking the brain, vol. I: Coding, vol. II: Consciousness, Oxford University Press
Norton, D. John (2003), Einstein's Special Theory of Relativity and the Problems in the Electrodynamics of Moving Bodies that Led him to it. Prepared for Cambridge Companion to Einstein, M. Janssen and C. Lehner, eds., Cambridge University Press
O'Connor, Timothy and Wong, Hong-Yu (2005), Emergent properties. In: Zalta E N (ed.), The Stanford Encyclopedia of Philosophy, (Summer 2005 Edition), URL = http://plato.stanford.edu/archives/sum2005/entries/properties-emergent/
Penrose, Roger (2004), The Road to Reality. A complete Guide to the Laws of the Universe, Jonathan Cape London
Phillips, A. William; von der Malsburg, Christoph and Singer, Wolf (2010), Dynamic Coordination in Brain and Mind. In von der Malsburg, Christoph, Phillips A. William, and Singer, Wolf (2010), Dynamic Coordination in the Brain From Neurons to Mind, The MIT Press Cambridge, Massachusetts London, England
Piccinini, Gualtiero (2006), Computational Explanation in Neuroscience. Synthese, 153, pp. 343–353
Pârvu, Ilie (2004), Posibilitatea Experienţei, (in English: The Possibility of Experience), Politeia-SNSPA.
Pessoa, Luiz; Tootell, B. H. Roger and Ungerleider, G. Leslie (2008), Visual perception of objects. In Squire et al. Fundamental neuroscience, Third edition, Elsevier Inc.
Place, T. Ullin (1956), Is consciousness a brain process?. British Journal of Psychology, in (eds.) Brian Beakley and Peter Ludlow, The Philosophy of Mind, A Bradford Book, The MIT Press
Place, T. Ullin (1988), Thirty years on-is consciousness still a brain process?. Australasian Journal of Philosophy, 66/2, pp. 208–219
Piccinini, Gualtiero (2006), Computational explanation in neuroscience. Synthese, 153, pp. 343–353
Pippin, B. Robert (1997), Idealism As Modernism – Hegelian Variations, Cambridge University Press
Poldrack, A. Russell (2011), The future of fMRI in cognitive neuroscience. NeuroImage, doi:10.1016/j.neuroimage.2011.08.007.
Poldrack, A. Russell (2010), Interpreting Developmental Changes in Neuroimaging Signals. Human Brain Mapping, 31, pp. 872–878
Prigojine, Ilya (1992), De la Existenţă la Devenire, Timp şi complexitate în ştiinţele fizice, Editura Ştiinţifică, translation of From Being to Becoming, 1980, San Francisco
Prinz, J. Jesse (2006), Is the Mind Really Modular. In Robert J. Stainton (ed.), Contemporary Debates in Cognitive Science, Blackwell Publishing
Putnam, Hillary (1987), The Many Faces of Realism, Open Court, La Salle, Illinois
Putnam, Hillary (2005), A Philosopher Looks at Quantum Mechanics (Again). British Journal of Philosophy of Science, 56, pp. 615–634

Pylyshyn, Zenon (2003), Return of the Mental Image: Are There Really Pictures in the Brain?. Trends in Cognitive Sciences, 7/3, pp. 113–118
Pylyshyn, Zenon (1999), Is Vision Continuous With Cognition? The Case for Cognitive Impenetrability of Visual Perception. Behavioral and Brain Science, 22(3), pp. 341–65
Quine, V. O. Willard (1968), Ontological Relativity. The Journal of Philosophy, vol. LXV, no. 7, in Quine, W. V. in Ontological Relativity and Other Essays, New York, Columbia University Press
Raichle, E. Marchus (2006), The Brain's Dark Energy. Neuroscience, vol. 314, pp. 1249–1250
Raichle, E. Marcus (2011), Intrinsic Activity and Consciousness. In S. Dehaene and Y. Christen (eds.), Characterizing Consciousness: From Cognition to the Clinic?, Springer-Verlag Berlin Heidelberg
Raichle, E. Marcus and Mintun, A. Mark (2006), Brain work and brain imaging. Annu. Rev. Neurosci., 29, pp. 449–76
Raichle, E. Marcus and Snyder, Z. Abraham (2009), Intrinsic brain activity and consciousness. In S. Laureys & G. Tononi (eds.) The Neurology of Consciousness, pp. 81–88
Ramachandran, S. Vilayanur and Sandra, Blakeslee (1998), Phantoms in the Brain, William Morrow and Company Inc., New York
Reid, R. Clay and Usrey, W. Martin (2008), Vision in Squire et al., Fundamental Neuroscience, 3rd edition, pp. 637–659
Rockwell, Teed (2003), Eliminativism. In Dictionary of Philosophy, on web page: (ed.) Chris Eliasmith, http://philosophy.uwaterloo.ca/MindDict/
Rosenblum, Bruce and Kuttner, Fred (2006), Quantum enigma. Physics encounters consciousness, Oxford University Press
Schmidt, Thomas (2009), Perception: The Binding Problem and the Coherence of Perception. In William P. Banks (Editor-in-chief) Encyclopedia of consciousness, vol. 2, Elsevier Inc.
Searle, R. John (1980), Minds, brains and programs. The Behavioral and Brain Sciences, 3, pp. 417–457
Searle, R. John (1992), The Rediscovery of the Mind, MIT Press.
Searle, R. John (1995), Consciousness, the Brain and the Connection Principle: A Reply. Philosophy and Phenomenological Research, 55. pp. 217–232
Searle, R. John (1999), The Chinese Room. In Wilson, R. A. and F. Keil (eds.), The MIT Encyclopedia of the Cognitive Science, Cambridge, MIT Press
Seymour, Kiley; Clifford, W. G. Colin; Logothetis, K. Nikos and Bartels, Andreas (2009), The coding of color, motion, and their conjunction in the human visual cortex. Current Biology, 19, pp. 177–183
Skarda, A. Christine and Freeman, J. Walter (1987), How the brain make chaos in order to make sense of the world. Behavioral and Brain Sciences, 10, pp. 161–195
Silberstein, Michael and McGeever, John (1999), The search for ontological emergence. The Philosophical Quarterly, 49:145, pp. 182–200
Smart, J. J. C. (1962), Sensations and brain processes. In V. C. Chappell (ed.) The Philosophy of Mind, Englewood
Smart, J. J. C. (2004), The identity theory of mind. The Stanford Encyclopedia of Philosophy (Fall 2004 Edition), Edward N. Zalta (ed.),
URL = http://plato.stanford.edu/archives/fall2004/entries/mind-identity/
Smolensky, Paul (1988), On the proper treatment of connectionism. Brain and Behavioral Science, 11, pp. 1–74
Smolin, Lee (2006), A crisis in fundamental physics. The New York Academy of Sciences, Jan/Feb issue
Sporns, Olaf (2006),Good Information? It's Not All About the Brain. November 2006,
http://www.sciencedaily.com/releases/2006/10/061027081145.htm
Stansbury, E. Dustin; Naselaris, Thomas and Gallant, L. Jack (2013), Natural Scene Statistics Account for the Representation of Scene Categories in Human Visual Cortex. Neuron, 79, pp. 1025–1034
Tallon-Baudry, Catherine (2009), The roles of gamma-band oscillatory synchrony in human visual cognition. Frontiers in Bioscience, 14, pp. 321–332

Tallon-Baudry, Catherine (2010), Neural coordination and human cognition. In von der Malsburg, C., Phillips A. W., & Singer, W., Dynamic Coordination in the Brain From Neurons to Mind, The MIT Press Cambridge, Massachusetts London, England

Tegmark, Max (2014), Our Mathematical Universe – My Quest for the Ultimate Nature of Reality, Alfred D. Knopf

Tegmark, Max (2004), Parallel universes. In D. John Barrow, C. W. Paul Davies, L. Charles Harper: 2004, Science and Ultimate Reality: Quantum Theory, Cosmology and Complexity, Cambridge University Press

Tegmark, Max and Wheeler, John Archibald (2001), 100 years of quantum mysteries. Scientific American, 284/2, pp. 68–75

Terhesiu, Dalia and Vacariu, Gabriel (2002), Brain, mind and the perspective of the observer. Revue Roumanie de Philosophie, 46/1–2, pp. 195–208

Thelen, Esther and Smith, Linda (1994), A Dynamic Systems Approach to the Development of Cognition and Action, MIT Press, Cambridge, MA

Uttal, R. William (2011), Mind and Brain – A Critical Appraisal of Cognitive Neuroscience, The MIT Press, Cambridge, Massachusetts, London, England

Vacariu, Gabriel; Terhesiu, Dalia and Vacariu, Mihai (2001) Towards a very idea of representation. Synthese, 129/2, pp. 275–295

Vacariu, Gabriel and Terhesiu, Dalia (2002), Brain, mind and the role of the observer. In Philosophy of Consciousness and Cognitive Science, Angela Botez and Bogdan Popescu (Eds.), Cartea Românească

Vacariu, Gabriel (2005) Mind, brain and epistemologically different worlds. Synthese, 143/3, pp. 515–548

Vacariu, Gabriel (2007) Kant, philosophy in the last 100 years and an epistemologically different worlds perspective. Revue Roumanie de Philosophie, 51, pp. 143–176

Vacariu, Gabriel (2008), Epistemologically Different Worlds, University of Bucharest Press

Vacariu, Gabriel and Vacariu, Mihai (2010), Mind, life and matter in the hyperverse, University of Bucharest Press

Vacariu, Gabriel (2011), Being and the hyperverse, University of Bucharest Press

Vacariu, Gabriel (2012), Cognitive neuroscience and epistemologically different worlds, University of Bucharest Press

Vacariu, Gabriel and Vacariu, Mihai (2013), Troubles with cognitive neuroscience. Philosophia Scientiae, 17/2 (France), pp. 151–170

Vacariu, Gabriel (2014), More troubles with cognitive neuroscience. Einstein's theory of relativity and the hyperverse, University of Bucharest Press

Van Gelder, Tim (1995), What might cognition be if not computation?. Journal of Philosophy, 92, pp. 345–381

Van Gelder, Tim (1999), Defending the dynamical hypothesis. In W. Tschacher and J. P. Dauwalder (eds.), Dynamics, Synergetics, Autonomous Agents: Nonlinear Systems Approaches to Cognitive Psychology and Cognitive Science, World Scientific, Singapore

Van Gelder, Tim and Port, F. Robert (1995), It's about time: A perspective to dynamical system approach to cognition. In R. Port and T. van Gelder (eds.), Mind as Motion: Explorations in the Dynamics of Cognition, MIT Press, Cambridge, MA

Van Gulick, Robert (2001), Reduction, emergence and other recent options on the mind/body problem – A philosophic overview. Journal of Consciousness Studies, 8/9–10, pp. 1–34

Velik, Rosemarie (2010), From single neuron-firing to consciousness—Towards the true solution of the binding problem. Neuroscience and Biobehavioral Reviews, 34, pp. 993–1001

Von der Malsburg, Christoph, (1999), The what and why of binding: the modeler's perspective. Neuron, 24, pp. 95–104

Von der Malsburg, Christoph (2010), Coordination What It Is and Why We Need It. In Von der Malsburg, C., Phillips A. W., and Singer, W. (2010), Dynamic Coordination in the Brain From Neurons to Mind, The MIT Press Cambridge, Massachusetts London, England

Vul, Edward; Harris, Christine; Winkielman, Piotr and Pashler, Harold (2009), Puzzlingly High Correlations in fMRI Studies of Emotion, Personality, and Social Cognition. Perspectives on psychological science, vol. 4, nr. 3, pp. 274–290

Waxman, W. (1995), Kant on the Possibility of Thought: Universals without Language. Review of Metaphysics, 48/4, pp. 809–57

Wahl, Russell (1999), How can what I perceive be true?. In T. Sorrell (ed), Descartes Aldershot, England, Brookfield, Vt.: Ashgate

Wheeler, Michael (2005), Reconstructing the Cognitive World – The Next Step, Bradford Book, The MIT Press

Wheeler, Michael (2009), The Problem of Representation. In Shaun Gallager and Daniel Schmicking (eds.), Handbook of Phenomenology and Cognitive Science, Springer

Wheeler, Michael and Clark, Andy (1999), Genic representation: reconciling content and causal complexity. The British Journal for Philosophy of Science, 50(1), pp. 103–135

Whitney, David (2009), Neuroscience: Toward unbinding the binding problem. Current Biology, 19/6, pp. R251–R253

Wilson, Catherine (1976), The epistemological argument for mind-body distinctness. Nous, vol. X, 3–15, in John Cottingham, 1998, Descartes, Oxford, New York: Oxford University Press

Woit, Peter (2006), Not Even Wrong – The Failure of String Theory and the Continuing Challenge to Unify the Laws of Physics, Jonathan Cape, London

Woit, Peter (2007), The problem with physics. Cosmos, 16, pp. 48–56

Wolbers, Thomas; Klatzky, L. Roberta; Loomis, M. Jack; Wutte, G. Magdalena and Giudice, A. Nicholas (2011), Modality-independent coding of spatial layout in the human brain. Current Biology, 21, pp. 984–989

Wolfson, Richard (2000), Einstein's Relativity and the Quantum Revolution: Modern Physics for Non-Scientists (movie with 24 episodes and his book), The Teaching Company.

Kolibri 360 GmbH
Ferdinand-Jühlke-Straße 7
99095 Erfurt, Deutschland
produktsicherheit@kolibri360.de